# Cities in Practice

Izhar ul Haq

Published by Izhar ul Haq, 2025.

CITIES IN PRACTICE

**First edition. October 29, 2025.**



ISBN: 979-8232425432

Written by Izhar ul Haq.

# CITIES IN PRACTICE

Planning with Integrity, Ecology, and the Politics of Space

**IZHAR UL HAQ**
**Principal Urban Planner**
**Lahore, Pakistan**
**FIRST EDITION – 2025**

For permission requests, please contact the author at:

**planner.izhar@gmail.com**

**First Edition – 2025**

Cities in Practice: First Edition 2025

# About the Author

**IZHAR UL HAQ** is a Principal Urban Planner with over 35 years of experience in urban planning, governance, and strategic development across Pakistan's diverse terrains. Known for his terrain-sensitive and ethically grounded approach, he has led multidisciplinary teams in crafting resilient, inclusive, and ecologically responsive urban strategies.

His career includes extensive consultancy work with the **Asian Development Bank (ADB)**, **World Bank**, **UNDP**, and other government and private sector entities. These engagements have shaped policy frameworks, institutional reforms, and field-tested solutions that respond to the complex realities of urban governance. His portfolio spans **12 ADB projects**, where he served as Independent Consultant, Staff Consultant, and Firm-based Consultant, alongside strategic assignments with **World Bank (5)**, **UNDP (2)**, and **CIDA (1)**—delivering insights on urban resilience, capacity building, and development strategy.

Beyond consultancy, IZHAR UL HAQ is a prolific public communicator. His widely read social media reflections—over 200 posts on planning ethics, spatial politics, and operational clarity—have fostered vibrant discourse among practitioners, students, and citizens. These reflections form the backbone of his forthcoming book, *Cities in Practice: Planning with Integrity, Ecology, and the Politics of Space*—a modular, audience-ready manuscript that distills field experience into actionable frameworks.

His core expertise includes:

- **Strategic Urban & Regional Planning** – master planning, land-use policy, and smart growth strategies
- **Governance & Policy Reform** – institutional capacity building and participatory governance
- **Development Financing & Grants** – crafting winning proposals for infrastructure and equity
- **Impact Evaluation & Analytics** – assessing project effectiveness through data-driven insights
- **Technical Communication** – persuasive reporting for stakeholders

at all levels

IZHAR UL HAQ continues to bridge policy and practice, aligning solutions with the evolving needs of cities and communities. His commitment to editorial clarity, visual storytelling, and operational alignment shapes how urban planning is taught, practiced, and understood—across platforms, disciplines, and generations.

# Chapter 1: Planning with Integrity

## 1.1 Planning as a Moral Practice

Urban planning is often misunderstood as a purely technical discipline—concerned with zoning codes, infrastructure grids, and regulatory compliance. In reality, it is a profoundly human endeavor. It shapes the spaces where people live, work, move, and dream. It determines access to opportunity, exposure to risk, and the quality of daily life. At its best, urban planning is a moral practice—one that demands clarity, substance, and integrity.

This book begins with a principle that has guided my professional journey: planning must be purposeful. It must be rooted in ethics, responsive to context, and committed to equity. The chapters that follow explore a wide range of urban challenges—from climate resilience and eco-tourism to spatial justice and disruptive technologies. But before we dive into those themes, we must first understand the foundation on which they rest: the planner's responsibility to act with integrity.

Planning is not merely about drawing maps or calculating densities. It is about making decisions that affect lives—decisions that must be made with care, foresight, and a deep sense of responsibility. Whether we are designing a master plan for a tourist corridor or enforcing zoning regulations in a growing city, our work must reflect the values we hold. These values—fairness, transparency, and inclusion—are not optional. They are the bedrock of ethical planning. These values have shaped my own journey, as the next section will show.

Planning is not a technical exercise—it is a civic ethic. Every zoning code, master plan, and infrastructure rollout encodes values. These values shape access, risk, and opportunity. Ethical planning demands clarity of purpose, transparency in process, and accountability in outcome.

When planners act with integrity, they do more than design cities—they steward public trust.

## 1.2 A Life in Planning

My work has spanned decades and disciplines. I have engaged with diverse aspects of Urban and Regional Planning—ranging from the preparation of local neighbourhood plans to the design of multi-metropolitan regional strategies. These experiences have included collaborations with governments, NGOs, and private sector partners to shape cities that are not only functional but also fair. I have authored technical guides, translated policy frameworks, and developed manuscripts that bridge local relevance with global accessibility.

Yet more than any title or credential, what defines my approach is editorial assertiveness. I believe in honoring the full scope of a project—no shortcuts, no condensation. When collaborators fall short, I demand a redo. When systems fail to serve the public, I challenge them. This assertiveness is not about ego; it is about accountability. It is about ensuring that every plan, every policy, and every page reflects the depth and dignity of the communities it serves.

This book itself is a product of that philosophy. It was not written in haste, nor was it assembled from fragments. Each entry was developed with care, tagged for thematic relevance, and reintegrated to preserve its original voice. The manuscript reflects a commitment to clarity, depth, and professional standards—values that every planner must uphold.

## 1.3 Ethics in Practice

Professional ethics are not an afterthought—they are the backbone of planning. In consultancy, ethical conduct means transparency with clients, honesty in assessments, and fairness in negotiations. It means resisting the temptation to rubber-stamp approvals or gloss over inconvenient truths. It means asking hard questions: Who benefits from this plan? Who is left out? What are the long-term consequences?

Ethics also shape how we communicate. I have always prioritized clarity—whether in technical reports, stakeholder presentations, or manuscript entries. Visual communication is a powerful tool, but it must be used responsibly. Infographics should illuminate, not obscure. Diagrams should clarify, not distract. And manuscripts must be structured with precision—each entry tagged, formatted, and reintegrated to preserve voice and substance.

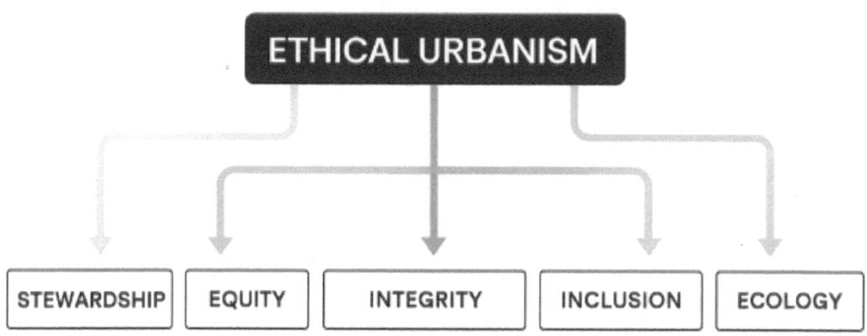


*Figure 1: Core principles of ethical urbanism.*

In my consultancy work, I have often encountered situations where ethical clarity was the difference between a successful project and a failed one. For example, in zoning enforcement, the temptation to overlook violations for political convenience is real. But doing so undermines the integrity of the entire planning system. Similarly, in regional development, ignoring environmental sensitivities for short-term gains can lead to irreversible damage. Ethics must guide not only what we plan, but how we plan—and for whom.

## 1.4 The Modular Mindset

One of the defining features of my workflow is modular manuscript development. Rather than writing in linear chapters, I build entries as discrete modules—each focused on a specific theme, case study, or concept. These modules are tagged for cross-referencing, organized for thematic coherence, and formatted for professional publication.

This approach mirrors the adaptive nature of planning itself. Cities are not static; they evolve. Plans must be updated, policies revised, and strategies reimagined. A modular mindset allows for flexibility without sacrificing depth. It enables planners to respond to emerging challenges—climate change, migration, technological disruption—while maintaining a coherent vision.

Modular development also supports collaborative workflows. Entries can be shared, reviewed, and refined without losing context. They can be translated, expanded, or merged to suit different audiences. Most importantly, they allow for iterative improvement—a principle that is as vital in writing as it is in planning.

Modularity is not just a formatting choice—it is a planning ethic. It reflects a commitment to clarity, adaptability, and operational alignment. It enables planners to navigate complexity without losing coherence, and to communicate with precision across diverse audiences and terrains.

## 1.5 Planning in Context

Context matters. One of the most common mistakes in urban development is the blind adoption of foreign models. Smart cities, for example, are often defined by high-tech solutions—sensor networks, AI-driven traffic systems, and digital governance platforms. These may work in advanced economies, but in many Pakistani cities, basic services are still lacking. We must define our own version of smartness—one that prioritizes equity, accessibility, and cultural relevance.

Similarly, eco-tourism must be rooted in local landscapes and traditions. Master plans should reflect the ecological character of a region, the aspirations of its communities, and the realities of its governance structures. Zoning enforcement must be sensitive to informal economies, spatial analysis must account for climate vulnerabilities, and stakeholder engagement must be inclusive of marginalized voices.

This is why I have always emphasized climate-responsive urban planning. It is not enough to build infrastructure—we must build resilience. We must design cities that can withstand heatwaves, floods, and resource scarcity. We must integrate nature-based solutions, promote adaptive land use, and ensure that planning decisions are informed by environmental data.

Planning in context is not a constraint—it is a compass. It ensures that our tools, language, and priorities are aligned with the lived realities of the places we serve.

To apply these principles meaningfully, we must confront the realities of planning practice in Pakistan—a terrain shaped by improvisation, inherited models, and civic urgency.

Urban planning is not a linear process—it is a negotiation between competing demands. A recent design reflection explores how planners must navigate trade-offs between density and livability, infrastructure and ecology, growth and equity. These tensions are not obstacles—they are the terrain of ethical planning. Context-sensitive decisions require clarity, courage, and civic alignment.

Smartness must be reframed as contextual planning. When digital infrastructure is deployed without civic alignment, it risks reinforcing exclusion. True smartness integrates technology with terrain, governance, and lived experience. A visual anchor reinforces this reframing.

## 1.6 Rethinking Urban Planning in Pakistan

Urban planning in Pakistan is often framed as a technical exercise—focused on master plans, zoning codes, and infrastructure grids. Yet this framing obscures the civic, ecological, and ethical dimensions of planning. A recent public reflection I shared challenges this view and argues that planning must be rethought as a civic practice—one that responds to terrain, governance, and lived experience.

This reflection emerged from decades of fieldwork and public discourse. It highlights three critical shifts:

- **From blueprint to dialogue**: Planning must move beyond static documents and embrace iterative, participatory processes.
- **From control to care**: Zoning and regulation must be tools of inclusion, not exclusion—protecting vulnerable communities and ecological assets.
- **From abstraction to alignment**: Planning language must be demystified, and planning tools must align with operational realities on the ground.

In Pakistan, urban planning frequently shifts between externally borrowed frameworks and context-driven improvisation. This hybrid practice can be fertile—but only if it is interrogated. We must ask: Who defines the planning agenda? Whose priorities are reflected? What values are embedded in our maps, overlays, and codes?

This reflection calls for planners to reclaim their role—not as technicians, but as civic stewards. It urges us to rethink planning not as a product, but as a process. Not as a document, but as a dialogue. And not as a control mechanism, but as a commitment to justice.

# 1.7 Planning Theory and the Evolution of Urban Practice

Urban planning theory is not an academic luxury—it is a living framework that shapes how cities are imagined, governed, and experienced. A recent public reflection I shared revisits this evolution, not to trace textbook milestones, but to illuminate how theory informs practice in real-world contexts.

Planning evolves through crises, reforms, and civic demands. From colonial control to post-war reconstruction, from modernist grids to participatory overlays, each era leaves behind tools, tensions, and lessons. In Pakistan, planning often oscillates between imported models and local improvisation. The result is a hybrid practice—part technocratic, part intuitive.

Yet theory matters. It helps us name what we do, critique what we inherit, and imagine what we must become. It allows planners to engage with foundational texts, map transitions, and contextualize their own practice within global and regional trajectories.

This reflection calls for planners to engage with theory not as dogma, but as dialogue. Because urban planning is not just about what we build—it is about how we think. And thinking clearly is the first step toward acting ethically.

## 1.8 Demystifying Planning Language

Urban planning is often wrapped in jargon—terms like "mixed-use," "TOD," "resilience overlays," and "spatial equity" circulate freely in reports and presentations. But for many citizens, these words obscure rather than clarify. A recent public reflection I shared challenges planners to rethink how we communicate.

Planning language should not be a gatekeeper—it should be a bridge. When we speak in acronyms and abstractions, we risk alienating the very communities we aim to serve. Technical precision is important, but so is accessibility, empathy, and narrative clarity.

This reflection calls for planners to adopt plain-language glossaries, visual storytelling, and community-tested framing. It urges us to translate complexity into clarity—not by dumbing down, but by opening up. Every zoning term, every policy tool, every design concept must be explainable in everyday language.

Planning is not just about what we build—it is about how we speak. And if our words cannot be understood, our plans cannot be trusted.

## 1.9 Governance as Planning Infrastructure

Governance is not a backdrop to planning—it is its operating system. Ethical planning requires more than technical skill or visionary design; it demands structural accountability. In Pakistan and beyond, planning decisions are often shaped by opaque processes, fragmented mandates, and short-term political pressures. This must change.

Governance must be reframed as a planning infrastructure. This means:

**Mandate clarity**: Planning authorities must be empowered with coherent jurisdiction, not diluted across departments.

**Process transparency**: Public hearings, zoning approvals, and procurement must be documented, accessible, and auditable.

**Outcome accountability**: Plans must be evaluated not just by deliverables, but by their impact on equity, resilience, and public trust.

**Data integration**: Health and climate data must inform zoning, overlays, and investment priorities.

**Civic voice**: Community input must be embedded in budgeting, procurement, and project evaluation.

**Transparency tools**: Dashboards, audit trails, and public feedback loops must be standard—not optional.

Accountability is not a compliance checkbox—it is a civic ethic. It requires planners to disclose assumptions, justify trade-offs, and engage with dissent. It demands that consultants challenge flawed terms of reference, that officials resist expedient approvals, and that citizens are empowered to shape their environments.

Governance reform is not a separate agenda—it is embedded in every master plan, zoning code, and infrastructure rollout. When planners treat governance as external, they risk becoming complicit in dysfunction. When they engage with governance as a design challenge, they become agents of institutional integrity.

Planning must be reframed as civic infrastructure—one that aligns spatial decisions with ethical governance. Because when governance fails, planning becomes performative. And when governance is principled, planning becomes transformative.

Lahore's future governance model emphasizes mandate clarity, civic dashboards, and outcome-based planning. These elements reinforce the planner's role as a steward of institutional integrity—not a technician of fragmented mandates. A visual anchor reinforces this alignment.

## 1.10 Temporal Urbanism: Planning with Time as Infrastructure

Urban planning often treats time as a constraint—deadlines, phasing schedules, and project lifecycles. But a recent public reflection I shared reframes time as a design material. Temporal urbanism asks: how do cities behave across hours, seasons, and decades? How do we plan not just for space, but for rhythm?

Time must be treated as infrastructure. This means:

• Chrono-zoning: Land use that adapts by hour or season—e.g., night markets, weekend plazas, monsoon buffers.

• Temporal overlays: Planning tools that map usage intensity, vulnerability, or accessibility across time.

• Rhythmic governance: Policies that respond to civic rhythms—school hours, prayer times, heatwaves, and migration cycles.

In Pakistan, where urban life is shaped by climate, culture, and informal economies, temporal planning is not optional—it is essential. A shaded corridor may be vital at noon but irrelevant at dusk. A street may host vendors by day and prayer overflow by night. Planning must reflect these shifts—not as exceptions, but as design logic.

Temporal urbanism is not about speed—it is about alignment. It is about designing cities that move with their people, respond to their rhythms, and evolve with their needs.

## 1.11 Myths and Truths About Urban Planning

Urban planning is often misunderstood—reduced to maps, codes, or technical jargon. These misconceptions distort public understanding and weaken civic engagement. A recent public reflection I shared challenges these myths and reframes planning as a civic ethic.

Planning is not neutral—it encodes values, priorities, and exclusions. It is not just technical—it is political, emotional, and cultural. It is not a one-time act—it is iterative, adaptive, and accountable. And it is not for planners alone—it must be co-created with communities.

Ethical planning begins by naming these truths. It demands that we demystify our tools, clarify our language, and confront our assumptions. Because only when planning is understood can it be trusted. And only when it is trusted can it serve.

## 1.12 Ethical Collaboration and Civic Engagement

Planning is not a solo act—it is a civic collaboration. Whether working with consultants, government officials, or community members, ethical engagement is essential. A recent public reflection I shared reframes collaboration as civic infrastructure: a system of trust, accountability, and shared purpose.

Professional partnerships must be grounded in clarity. Terms of reference must be explicit. Deliverables must be audience-ready. Feedback must be constructive, not performative. I have often mentored collaborators—challenging them to refine their work, clarify their visuals, and align their outputs with public impact. This is not micromanagement—it is stewardship.

Community engagement must go beyond tokenism. Public consultations must be designed for inclusion, not optics. Marginalized voices must be centered, not sidelined. Civic knowledge—local histories, informal networks, lived experience—must be treated as planning data. Because without community insight, planning becomes extractive.

Ethical collaboration is not just about working together—it is about working responsibly. It is about ensuring that every stakeholder, from consultant to citizen, is treated with respect, clarity, and care. And it is about recognizing that civic engagement is not a checkbox—it is the foundation of ethical planning.

## 1.13 Small Cities, Big Futures

Urban planning discourse often privileges megacities—Karachi, Lahore, Islamabad—as the primary sites of innovation and investment. But this framing overlooks the transformative potential of small and intermediate cities. A recent public reflection I shared reframes these cities not as peripheral, but as pivotal.

Small cities are often more agile. Their governance structures are less encumbered, their communities more cohesive, and their spatial challenges more manageable. They offer unique opportunities for climate-responsive planning, inclusive zoning, and participatory design. They are not testing grounds—they are futures in motion.

This reflection calls for planners to shift their lens. Instead of scaling down metropolitan models, we must scale up local ingenuity. We must recognize that small cities are not miniature versions of big ones—they are distinct ecosystems with their own rhythms, risks, and resilience strategies.

Planning must move beyond hierarchy. It must embrace diversity—not just of scale, but of possibility. Because the future of urban Pakistan will not be built in one city—it will be co-authored across many.

Revitalising small cities is not a nostalgic gesture—it is a strategic imperative. A recent design reflection explores how targeted investments, civic infrastructure, and terrain-sensitive planning can unlock the latent potential of Pakistan's intermediate cities. These urban nodes are not waiting rooms for metropolitan migration—they are laboratories of resilience, equity, and innovation. Planning must recognize their agency and design accordingly.

## 1.14 Master Planning in Pakistan

Master planning in Pakistan is often treated as a ceremonial exercise—commissioned, compiled, and shelved. A recent public reflection I shared critiques this pattern and calls for a redefinition of what master planning should mean in our context.

Too often, master plans are disconnected from governance, budgeting, and civic engagement. They are drafted without operational clarity, approved without public scrutiny, and implemented without accountability. This is not planning—it is performance.

A meaningful master plan must be:

- Operationally aligned: Integrated with zoning, budgeting, and institutional mandates.
- Civically anchored: Developed through inclusive consultations and public review.
- Climate-responsive: Informed by environmental data, risk assessments, and resilience strategies.
- Modular and adaptive: Designed for iteration, not rigidity—able to evolve with terrain, governance, and civic demand.

This reflection urges planners to reject ceremonial planning and embrace substantive reform. It calls for master plans that are not just documents, but commitments. Not just maps, but mandates. And not just visions, but vehicles for ethical urban transformation

Planning is not just a technical act—it is a moral one. But morality must extend beyond people to place. The next chapter explores how terrain is erased, ecology is forgotten, and planning must learn to see what it routinely overlooks.

# Chapter 2: Vanishing Ecologies

## 2.1 Planning with Terrain, Memory, and Urgency

If planning is a moral practice, then ecology is its first test. This section reframes the disappearance of wetlands, forests, floodplains, and biodiversity corridors—not as environmental losses, but as planning failures. These systems are not passive victims of urban growth; they are casualties of zoning codes, plot grids, and governance blind spots.

In Pakistan, ecological erasure is systemic. Rivers are renamed as drains. Wetlands are omitted from master plans. Contours are flattened into cadastral geometry. This is not just technical oversight—it is ethical neglect.

Planning must begin with ecological memory: the ability to map what once existed, understand what remains, and imagine what can be restored. Terrain is not an obstacle—it is infrastructure. It regulates climate, absorbs shocks, and anchors civic identity. When planners ignore terrain, they invite disaster. When they honor it, they build resilience.

This section calls for terrain-based planning that is:

- Responsive to slope, soil, and seasonal logic
- Informed by hydrological thresholds and biodiversity corridors
- Aligned with climate data and civic rhythms
- Designed for restoration, not just mitigation

Vanishing ecologies are not just environmental concerns—they are ethical indictments. They reveal what planning has ignored, and what it must now protect. Because without ecological integrity, urban resilience is a myth. And without terrain literacy, planning becomes complicit in collapse.

Vanishing ecologies are not accidents—they are outcomes of planning choices. When terrain is ignored, memory erased, and urgency deferred, cities lose more than biodiversity—they lose resilience, identity, and justice. Ethical planning must treat ecology as infrastructure, not scenery. Planners must ask: What are we erasing? And who decides what survives?

## 2.2 Planning Blind Spots in Air Quality Governance

Air pollution in Pakistan is not just a seasonal crisis—it is a structural failure. The smog season reveals how deeply disconnected planning is from environmental governance. Despite recurring public health emergencies, master plans and zoning codes remain silent on air quality thresholds, pollution overlays, and ecological buffers.

This silence is not accidental. It reflects a planning culture that treats air as invisible, and pollution as someone else's problem. Mandates are fragmented across departments. Enforcement is delayed or diluted. Data is denied, disputed, or buried. And civic pressure is deflected with technical jargon and bureaucratic opacity.

Smog is not evenly distributed. It settles in low-income neighborhoods, traffic corridors, and industrial zones—places with weak ventilation, poor tree cover, and little political leverage. This is **urban injustice**: where pollution maps overlap with poverty maps, and clean air becomes a privilege.

The **maths of air pollution** are brutal. Lahore's AQI routinely exceeds 400 during peak smog, with PM2.5 levels up to 30 times higher than WHO thresholds. These are not abstract numbers—they translate into asthma attacks, missed school days, lost wages, and premature deaths. Yet planning documents rarely cite them.

Occasionally, the smog clears—due to wind shifts, rain, or regional anomalies. These **atmospheric anomalies** confuse the public and dilute civic pressure. But they do not solve the problem. They merely postpone it.

Smog is also **regional**. Lahore's air is shaped by cross-border crop burning, regional wind patterns, and transboundary industrial emissions. Yet planning remains city-bound, ignoring the ecological scale of the crisis.

During peak smog, Lahore experiences **atmospheric breakdown**. Visibility drops. Schools close. Hospitals overflow. Civic life stalls. And yet, planning logic remains unchanged—no overlays, no thresholds, no accountability.

Planning blind spots include:

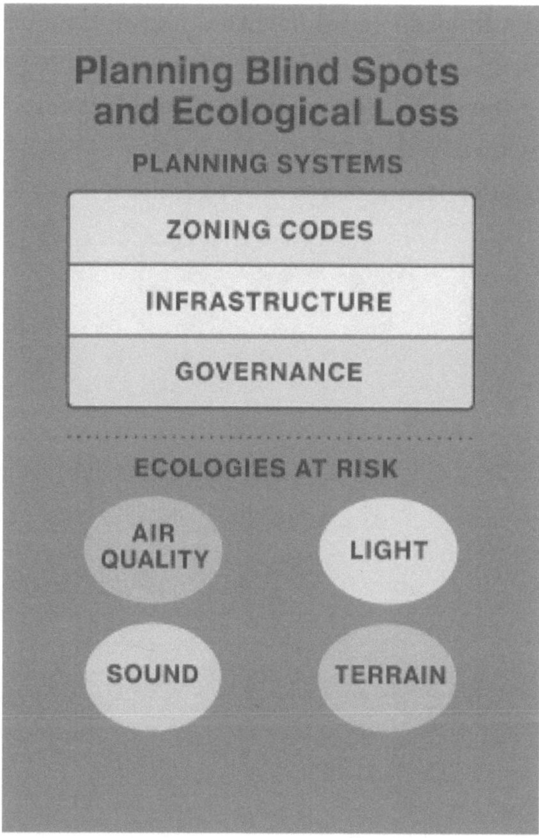


- **No overlays for pollution-sensitive zones**: Industrial clusters, traffic corridors, and low-ventilation basins are rarely mapped or regulated.
- **No air quality thresholds in zoning or building codes**: Planning instruments ignore particulate matter, wind channels, and seasonal inversion risks.
- **No ecological buffers in urban expansion**: Wetlands, tree corridors, and open spaces are erased or encroached without consequence.
- **No integration with public health data**: Hospitals and schools are sited without reference to pollution exposure or respiratory risk.
- **No civic accountability mechanisms**: Citizens cannot trace pollution sources, challenge approvals, or demand mitigation through planning channels.

This section calls for a fundamental shift: from passive documentation to active regulation. Air quality must be embedded in planning logic—not as an afterthought, but as a threshold. Because when planning ignores the air, it abandons the people who breathe it.

Beyond the air we breathe lies another vanishing frontier—the sky itself.

## 2.3 We Gained the World and Lost the Sky

Urban expansion has not only erased wetlands and forests—it has stolen the sky. In cities like Lahore, the stars are gone. The Milky Way is invisible. The night is no longer a canvas of wonder, but a ceiling of haze and glare. This loss is not poetic—it is political. It reflects a planning logic that prioritizes brightness over balance, and visibility over vision.

We gained the world—roads, towers, malls, and flyovers. But we lost the sky. Children grow up without constellations. Elders forget the rhythm of moonrise. Civic rituals tied to the heavens—Ramadan moons, solstice prayers, seasonal migrations—are now mediated by screens, not by stars.

This is not just cultural erosion—it is ecological blindness. The sky is a living system. It regulates temperature, anchors biodiversity, and connects communities across time and terrain. Its disappearance signals a deeper disorientation: where planning no longer respects the vertical dimension of urban life.

Planning must reclaim the sky as a civic right. This means:

- Recognizing light pollution as a planning failure, not just a technical nuisance.
- Embedding sky visibility into urban design, overlays, and building codes.
- Protecting dark corridors and celestial sightlines, especially in public spaces.
- Restoring the night as a shared experience, not a privatized glare.

The disappearance of stars dims our civic imagination—shrinking our sense of scale, wonder, and shared identity. When planning excludes the sky, it excludes the people who dwell below.

## 2.4 Artificial Light and the Erasure of Stars

Light pollution is the most overlooked form of urban excess. It is silent, constant, and normalized. In cities like Lahore, artificial light floods every surface—billboards, flyovers, gated colonies, and commercial strips. The night is no longer dark. It is over lit, overdesigned, and overregulated by brightness.

This erasure of stars is not just aesthetic—it is ecological. Artificial light disrupts migratory birds, confuses pollinators, and fragments nocturnal habitats. It also disorients humans. Sleep cycles are disturbed. Mental health suffers. And civic rituals tied to the night sky—moon sightings, solstice gatherings, and stargazing—fade into memory.

Planning documents rarely mention light pollution. Building codes regulate height and setbacks but not glare. Zoning maps ignore celestial corridors. Public lighting is designed for visibility, not balance. And commercial lighting is treated as a sign of progress, not a source of harm.

This section calls for a new planning ethic—one that respects darkness as a civic and ecological asset. Planners must:

**Map light pollution zones**, especially near ecological corridors and residential areas.

**Regulate commercial lighting**, billboards, and façade glare through overlays and building codes.

**Protect dark corridors**, especially in parks, wetlands, and cultural sites.

**Design lighting for rhythm**, not just reach—aligning with seasonal cycles and civic needs.

Because when the stars disappear, so does our sense of scale, wonder, and ecological belonging. And when planning forgets the night, it forgets half the city.

As I reflected in one of my public posts, city lights shine bright—but at what cost? The sky mourns its missing stars. This loss is not just ecological—it is emotional, cultural, and civic. When we erase the night sky, we erase wonder, memory, and orientation. Planning must protect darkness as much as it provides light.

Light poverty is not just the absence of stars—it is the erasure of civic orientation, ecological rhythm, and emotional scale. When planning floods the night with glare, it privatizes visibility and fragments nocturnal habitats. Children lose constellations. Elders lose moonrise. Cities lose memory. Ethical planning must protect darkness as infrastructure—not just regulate brightness. Because when the stars disappear, so does our sense of belonging.

## 2.5 Acoustic Urbanism — Listening as Planning Infrastructure

Sound is not ambient—it is civic. In cities across Pakistan, soundscapes reveal rhythm, identity, and exclusion. From mosque loudspeakers and traffic horns to construction clatter and informal street music, urban sound tells us who is heard, who is silenced, and how space is experienced.

This section introduces **acoustic urbanism**—the idea that sound must be treated as infrastructure. Acoustic environments shape mental health, social cohesion, and ecological behavior. Yet planning rarely accounts for sonic equity. Noise bylaws are outdated. Decibel audits are rare. And sound-sensitive design is dismissed as aesthetic, not operational.

Ethical planning demands that we listen deeply. This means:

**Sound audits** that map decibel levels, frequency patterns, and vulnerable zones.

**Zoning overlays** that regulate acoustic thresholds—especially near schools, hospitals, and green spaces.

**Design strategies** that absorb, redirect, or celebrate sound—through materials, layout, and civic programming.

In Pakistan, where informal economies and religious practices shape urban acoustics, planners must listen with care. A street vendor's call may be disruptive to some, but vital to others. A quiet park may offer refuge, but also erase cultural vibrancy.

Acoustic urbanism is not about silence—it is about alignment. It asks planners to hear the city as lived experience, not just technical noise. Because sound is not just heard—it is inhabited.

## 2.6 Solar Justice — Protecting Sunlight in Vertical Cities

Sunlight is not just a natural resource—it is a civic right. As Pakistan's cities expand vertically, urban policies must ensure that high-rise development does not deprive low-rise buildings and marginalized communities of sunlight. The loss of sunlight creates **light poverty**—a condition that deepens inequality, worsens public health, and undermines urban resilience.

In cities like Lahore, towering structures often cast prolonged shadows over poorer neighborhoods. These shadows are not neutral—they are socio-environmental disparities. Sunlight is essential for physical health, mental well-being, solar energy, and ecological vitality. Its absence contributes to vitamin D deficiency, seasonal depression, reduced solar panel efficiency, and the decline of urban greenery.

Shade is vital—but relying on skyscrapers for cooling relief is inadequate. Ethical planning must offer alternatives: tree-lined streets, pergolas, jaali screens, and reflective surfaces that cool without eclipsing. **Shadow equity** must be embedded in planning frameworks.

This section calls for:

**Shadow impact assessments** before approving high-rise projects.

**Sunlight clauses in building codes** to guarantee minimum exposure, especially in winter.

**Zoning overlays** that distinguish sunny zones (e.g., parks, hospitals, solar farms) from shaded zones (e.g., markets, transit corridors).

**Solar rights protections** for rooftop panels and informal settlements.

**Community participation** in development approvals affecting sunlight access.

**Seasonal shadow laws** that adapt to Pakistan's climate rhythms.

**Smart shading strategies** using trees, screens, and architectural gaps to balance shade and light.

Solar justice is not anti-density—it is pro-equity. It calls for a contextual approach that integrates shadow analysis into urban planning and protects sunlight for the vulnerable. Because when sunlight is privatized, inequality is cast in shadow.

## 2.7 Lost in Smog: Lahore's Atmospheric Breakdown

Lahore's smog crisis is not a seasonal nuisance—it is a civic indictment. Each winter, the city's air becomes unbreathable, revealing the fragility of its planning systems and the erosion of public trust. The sky stings. The horizon disappears. And the baseline for livability collapses.

Smog is not a meteorological condition—it is a governance failure. It reflects:

**Policy inertia**: Emission controls remain unenforced. Transit reform is delayed. Industrial emissions go unmonitored.

**Spatial negligence**: Industrial zoning overlaps with residential life. Ecological buffers are erased. Ventilation corridors are ignored.

**Civic disempowerment**: Citizens are expected to adapt, not advocate. Masks replace mandates. Apps replace accountability.

The smog season now dictates school closures, hospital surges, and outdoor avoidance. But behind these symptoms lies a deeper erosion: the normalization of toxicity. When children grow up expecting the sky to sting, the city has failed.

Ethical planning demands:

**Air quality as infrastructure**—integrated into zoning, mobility, and investment decisions.

**Public health audits**—to track the spatial distribution of respiratory risk and exposure.

**Civic mobilization**—to shift the narrative from tolerance to transformation.

This section positions smog not as weather, but as warning. Because when the air becomes unlivable, the city becomes ungovernable.

# 2.8 Smog Without Borders — Lahore's Regional Crisis

Lahore's smog is not a local anomaly—it is a regional indictment. Each winter, the haze that blankets the city is shaped not only by urban congestion, but by transboundary agricultural burn-off, industrial emissions, and meteorological patterns that ignore administrative borders. The air does not respect jurisdictional lines. Neither should planning.

This is not just an environmental failure—it is a governance failure. Emissions drift across districts. Burn cycles are misaligned. Wind patterns carry pollutants from one province to another. Yet planning remains city-bound, fragmented, and reactive.

Ethical planning demands:

- **Regional coordination** between Punjab, Sindh, and cross-border jurisdictions to align burn cycles, emission controls, and air quality thresholds.
- **Meteorological integration** in planning overlays—recognizing how wind, inversion layers, and seasonal shifts affect pollution spread.
- **Public transparency** in pollution data, enforcement actions, and mitigation strategies—so citizens can trace accountability across borders.

Occasionally, the smog clears—prompting relief, but also reflection. Was it policy, weather, or coincidence? Clean air must be understood, not just celebrated. Ethical planning demands systemic inquiry into atmospheric shifts, equitable distribution of air quality benefits, and replicable strategies rooted in governance—not luck. Because clean air must be a standard, not a surprise.

Territorial resilience must replace city-scale interventions. Planning must evolve to reflect ecological boundaries—not just political ones. Because when the air is shared, so is the responsibility.

But ecological resilience is not only regional—it is relational. It includes those who inhabit the city beyond the human.

## 2.9 Non-Human Agency — Rethinking Urban Life Beyond the Human

Urban planning in Pakistan remains deeply anthropocentric. Cities are imagined as human-only domains—designed for movement, consumption, and control. Yet non-human actors—animals, plants, microbes, and atmospheric forces—shape urban rhythms, risks, and resilience in ways that planning rarely acknowledges.

Stray dogs regulate street dynamics. Birds signal ecological health. Trees moderate temperature. Microbial ecosystems influence sanitation and public health. These entities are not passive—they interact, adapt, and co-create urban experience. Their presence reveals the permeability, adaptability, and vitality of urban space.

Ethical urbanism must recognize **non-human agency** as part of civic life. Planning must move beyond human-centered metrics and embrace **multispecies urbanism**. This means:

**Ecological zoning** that protects habitats and migratory corridors.

**Design strategies** that accommodate animal movement, nesting, and refuge.

**Policy frameworks** that treat biodiversity as infrastructure—not ornament.

**Public education** that fosters respect for non-human life in urban settings.

In Pakistan, informal settlements, green belts, and peri-urban zones host rich biodiversity. Planners must engage with these ecologies—not erase them. The presence of non-humans is not a nuisance—it is a signal. It challenges planners to ask: Who else lives here? Who else shapes this space? And how can we plan with—not just for—them?

Non-human agency reframes planning as a shared negotiation—not just between stakeholders, but between species. It expands the civic imagination and deepens the ethical field. Because cities are not only built—they are inhabited.

## 2.10 Paradigm Shifts — Rethinking Urban Development Models

The dominant paradigms of urban development—those that prioritize growth over ecology, density over dignity, and infrastructure over intimacy—must be challenged. In Pakistan and similar contexts, imported models often fail to account for terrain, climate, and civic rhythm. The result is a disconnect between what is built and what is needed.

Urban development paradigms must be contextualized, not copied. Planning cannot be reduced to replication. It must be a form of translation—between global ideas and local realities, between technical frameworks and lived experience.

Ethical planning demands:

**Ecological alignment**: Development must respond to local climate, biodiversity, and resource cycles—not override them.

**Civic calibration**: Planning tools must reflect informal economies, cultural rhythms, and spatial memory—not erase them.

**Paradigm pluralism**: No single model fits all cities. Planners must curate, adapt, and interrogate frameworks—not adopt them wholesale.

Vanishing ecologies are not just environmental losses—they are paradigm failures. When development ignores context, it erases more than trees—it erases meaning. Planning must reclaim its role as a translator, not a technician. Because cities are not just built—they are interpreted.

Ecology is not just terrain—it is testimony. But testimony must be heard. The next chapter turns to participation: how power is distributed, how communities are engaged, and how planning must evolve from imposition to invitation.

# Chapter 3: Participation and Power

## 3.1 Who Plans the City?

Urban planning decisions are frequently made in boardrooms, municipal offices, and consultancy hubs—far removed from the communities they affect. While participatory planning is widely promoted, its practice often falls short. Public hearings are scheduled at inaccessible times, technical jargon dominates discussions, and community input is filtered through bureaucratic lenses. The result is a planning process that privileges expertise over experience, and institutional convenience over democratic inclusion.

This disconnect is not accidental. Historically, planning has served as a tool of control—used to regulate populations, protect property interests, and enforce social hierarchies. From colonial town layouts to modern-day redevelopment schemes, the voices of marginalized groups have been systematically excluded. Informal settlements, indigenous communities, and working-class neighborhoods are often treated as problems to be solved rather than partners in shaping urban futures.

Even when participation is invited, it is often tokenistic. Stakeholder consultations may involve a handful of pre-selected representatives, while broader community concerns are sidelined. Participatory mapping exercises may be used to validate pre-existing plans rather than generate new ones. The language of inclusion masks a reality of exclusion.

## 3.2 Structural Participation

To move beyond this, planners must embrace structural participation. This means redistributing decision-making power—not just collecting feedback, but co-creating solutions. Models such as community land trusts, participatory budgeting, and citizen planning assemblies offer pathways toward more democratic urban governance. These approaches recognize that lived

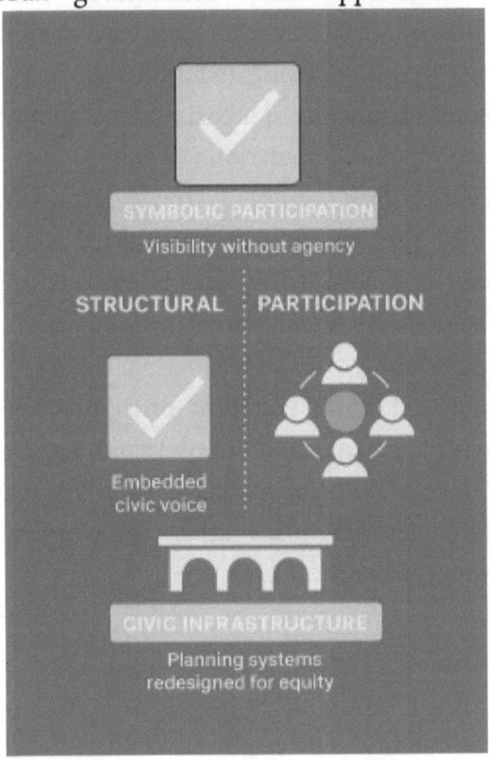


experience is a form of expertise, and that planning must be accountable to those it affects.

Moreover, participation must be intersectional. Gender, race, class, and ability all shape how people experience the city—and how they engage with planning processes. Inclusive planning must therefore be attentive to these differences, ensuring that engagement is not just broad, but deep.

Ultimately, the question is not whether participation is possible, but whether planners are willing to relinquish control. True participation requires humility, transparency, and a commitment to justice. It demands that we ask not just "what should be built?" but "who decides?" and "who belongs?"

Community participation is not a procedural formality—it is a planning infrastructure. When citizens co-design their environments, they shift planning from prescription to partnership. A visual anchor reinforces this civic imperative.

Participation is not a checkbox—it is civic infrastructure. When communities co-design their environments, they shift planning from prescription to partnership. Public hearings, mapping exercises, and stakeholder consultations must be more than rituals—they must redistribute power. Ethical planning demands that we listen not just to data, but to dissent. Because when participation is performative, planning becomes exclusion dressed as inclusion.

## 3.3 Partnerships as Planning Infrastructure

In urban planning, the strength of a project often hinges not on its technical brilliance, but on the integrity of its partnerships. A recent public reflection I shared emphasized that client-consultant relationships are not transactional—they are strategic infrastructures built on trust, clarity, and shared purpose.

Trust is earned through consistency, transparency, and a commitment to outcomes that serve both public interest and professional standards. Clarity, meanwhile, is operational: consultants must communicate not only through deliverables, but through how they frame problems, navigate constraints, and propose solutions. Shared purpose emerges when both parties align around a vision that transcends individual roles—whether crafting a master plan, reforming zoning codes, or designing climate-resilient infrastructure.

These principles are not abstract. They shape how planning unfolds in real time. When trust falters, projects stall. When clarity is compromised, misalignment grows. And when purpose is fragmented, outcomes lose coherence. Strong partnerships enable planners to act not merely as service providers, but as strategic collaborators—facilitating ethical, responsive, and community-aligned development.

This philosophy echoes the broader theme of participatory planning: that expertise must be paired with humility, and that planning is most effective when rooted in relational accountability. In this sense, partnerships are not just professional arrangements—they are civic commitments.

## 3.4 Reflection: Participation Without Power

In one consultancy engagement, I was invited to review a redevelopment proposal for a historic neighborhood. The plan included upgraded infrastructure, new zoning overlays, and a beautification strategy. On paper, it was progressive. But when I asked about community input, the response was vague: "We held a consultation." That consultation, it turned out, involved three pre-selected stakeholders, none of whom lived in the area.

Ecology is not just terrain—it is testimony. But testimony must be heard. The next chapter turns to participation: how power is distributed, how communities are engaged, and how planning must evolve from imposition to invitation.

I visited the neighborhood myself. Residents spoke of displacement fears, rising rents, and the erasure of cultural landmarks. They had not seen the plan. They had not been asked what mattered to them. Their lived experience—decades of adaptation, resilience, and informal stewardship—was absent from the planning logic.

This is not an isolated case. Across Pakistan, participation is often performative. It is used to validate decisions already made, not to shape them. Maps are drawn without memory. Codes are enforced without context. And communities are consulted only when resistance is anticipated.

Ethical planning demands more. It requires planners to show up—not just with frameworks, but with humility. It requires us to listen—not just to data, but to dissent. And it requires us to co-create—not just documents, but futures.

Participation without power is not inclusion—it is illusion. And planning that ignores this truth risks becoming a tool of exclusion dressed in democratic language.

Participation is not a slogan—it is a structure. But structures are shaped by rules. The next chapter examines zoning: how control is exercised, how space is regulated, and how planning must confront the invisible architectures of exclusion.

# Chapter 4: Zoning and Control

## 4.1 Zoning as a Language of Power

Zoning is often described as the skeleton of urban planning—a framework for organizing land use, managing growth, and protecting public welfare. Yet zoning is also a language of power. It encodes political priorities, shapes social outcomes, and reflects ethical choices about who gets to live where, and under what conditions.

At its core, zoning divides the city into categories: residential, commercial, industrial, recreational. These categories are not neutral—they are shaped by historical legacies, economic interests, and political agendas. Single-family zoning, for example, has been used to preserve affluent neighborhoods and exclude lower-income residents. Industrial zoning has often been concentrated near marginalized communities, exposing them to environmental hazards.

Zoning is not neutral—it is a language of power. Every overlay, setback, and land use designation encodes priorities: who belongs, who benefits, and who is displaced. When zoning is framed as technical, it hides its political and ethical consequences. Ethical planning demands that we treat zoning not as control, but as care—protecting ecological assets, informal livelihoods, and civic memory. Because when zoning is misused, it becomes infrastructure for exclusion.

## 4.2 Master Plans and Ethical Blind Spots

Master plans, which guide long-term development, carry similar ethical weight. They project visions of the future—often decades ahead—yet rarely reflect the aspirations of those most affected. These documents are typically drafted by consultants and approved by officials, with limited public input. They prioritize growth, investment, and aesthetics, while sidelining equity, affordability, and ecological resilience.

The ethical challenge of zoning lies in its dual role: it is both a technical tool and a moral instrument. It shapes not just buildings, but lives. It determines access to schools, parks, jobs, and transit. It influences property values, displacement risks, and community cohesion.

## 4.3 Zoning and the Ethics of Control

Zoning is not merely a regulatory framework—it is a system of control. It decides who belongs where, and under what terms. In many cities, zoning has been used to entrench inequality, segregate communities, and prioritize private interests over public good.

The ethical implications are profound. When zoning codes reinforce exclusionary patterns, they become instruments of injustice. When master plans ignore informal settlements or ecological fragility, they perpetuate harm. Ethical zoning demands more than compliance—it demands conscience.

Planners must interrogate the assumptions embedded in zoning categories. They must ask: Who benefits from this designation? Who is displaced? What alternatives exist? Ethical zoning is not about perfection—it is about accountability.

Equitable land use is not a technical achievement—it is a civic ethic. A recent design reflection explores how balanced spatial distribution influences city character and economic vitality. When zoning reinforces exclusion, it erodes trust and opportunity. When it enables access, it becomes a tool of justice. Planners must treat land use not as a neutral grid, but as a moral map.

## 4.4 Toward Responsive Zoning

Responsive zoning requires planners to move beyond static categories and embrace adaptive overlays. Climate-responsive zoning, for example, integrates environmental data into land use decisions. Social overlays prioritize affordable housing, inclusive mobility, and public health.

Technology can support this shift, but it cannot replace ethical judgment. GIS tools, zoning dashboards, and predictive models must be paired with community engagement, historical awareness, and ecological sensitivity.

Ultimately, zoning must evolve from a tool of control to a framework of care. It must reflect not just what is permissible, but what is just.

# 4.5 Logarithmic Zoning: A New Planning Logic

Traditional zoning frameworks often rely on linear thresholds—fixed plot sizes, uniform setbacks, and rigid land use categories. But cities are not linear. Their densities, risks, and opportunities scale unevenly. Logarithmic zoning offers a new logic: one that adjusts regulatory intensity based on exponential patterns of growth, vulnerability, or ecological stress.

For example, instead of applying the same setback rule across all zones, a logarithmic model might increase setbacks exponentially in flood-prone areas or reduce them in high-access corridors. This approach allows planners to respond to urban complexity with mathematical nuance—aligning regulation with lived realities.

Logarithmic zoning is not just a technical innovation—it is an ethical recalibration. It recognizes that spatial justice requires differentiated responses. It invites planners to move beyond one-size-fits-all codes and embrace a spectrum of regulatory sensitivity.

This concept is still evolving, but its potential is clear. It could inform climate overlays, mobility corridors, and housing affordability zones. It could reshape how we think about zoning not as a grid, but as a gradient.

Logarithmic zoning demands that we rethink not only categories, but metrics. If zoning is to reflect ethical density and ecological alignment, then its numerical backbone—Floor Area Ratio (FAR)—must be interrogated. FAR is often treated as a neutral planning tool, yet it encodes privilege, exclusion, and spatial inequity. Before we can recalibrate zoning logic, we must understand how FAR operates—and how it can be reimagined.

FAR is not just a metric—it is a spatial ideology. A recent design reflection explains how Floor Area Ratio shapes urban form, density, and access in Pakistan's cities. Whether embedded in zoning codes or negotiated through exceptions, FAR reflects power. Planners must interrogate its logic: Who benefits from higher FAR? Who is displaced by its limits? And how can we recalibrate FAR to serve equity, ecology, and urban resilience?

## 4.6 Snapshot: Zoning as Displacement Infrastructure

In a mid-sized city in Punjab, a master plan was approved that reclassified a mixed-use neighborhood into a commercial corridor. The justification was economic: increased footfall, retail growth, and transit connectivity. But the zoning shift triggered a cascade of displacement. Long-standing residents—many from working-class backgrounds—faced rising rents, noise pollution, and the loss of communal spaces. Informal vendors were evicted. Schools and clinics were relocated. The neighborhood's social fabric unraveled.

When planners were asked about community engagement, they cited a public notice and a consultation with commercial stakeholders. No participatory mapping was conducted. No displacement audit was performed. The zoning code had been applied uniformly—without regard for tenure patterns, cultural rhythms, or ecological fragility.

This is not an anomaly. Across Pakistan, zoning is often wielded as a tool of transformation without accountability. It reshapes space, but also erases memory. It enables investment but also displaces vulnerability. And when applied without ethical scrutiny, it becomes infrastructure for exclusion.

Responsive zoning demands more than technical overlays. It requires planners to ask: Who lives here? Who is at risk? And how can regulation protect—not displace—the communities it governs? Zoning is not always neutral—it can be a tool of displacement. A recent design reflection explores how planning decisions, when divorced from ethics, can uproot communities and erase informal resilience. From eviction drives to redevelopment schemes, displacement is often framed as progress. But planners must ask: progress for whom? Ethical zoning demands that we confront the human cost of spatial control.

## 4.7 Glossary Breakout: Logarithmic Zoning

Logarithmic zoning is a planning logic that replaces linear thresholds with exponential sensitivity. Traditional zoning frameworks often apply uniform rules—fixed plot sizes, standard setbacks, and rigid land use categories—regardless of ecological risk or urban intensity. But cities do not scale linearly. Their densities, vulnerabilities, and opportunities follow complex, often exponential patterns.

Logarithmic zoning responds to this complexity. It proposes that regulatory intensity should scale with risk, access, or ecological fragility. For example:

- In flood-prone zones, setbacks might increase exponentially with proximity to water bodies.
- In high-access corridors, density allowances might expand logarithmically to support transit-oriented development.
- In ecologically sensitive areas, land use permissions might tighten based on cumulative stress indicators.

This approach allows planners to calibrate regulation with nuance—aligning zoning codes with lived realities rather than abstract categories. It also introduces a mathematical ethic: that spatial justice requires differentiated responses, not one-size-fits-all solutions.

Logarithmic zoning is still evolving. It requires new data models, adaptive overlays, and civic engagement. But its promise lies in its ability to translate complexity into care—making zoning not just a tool of control, but a framework of responsiveness.

Zoning is not just about control—it is about consequence. And the greatest consequence of poor planning is climate vulnerability. The next chapter explores how cities must respond to rising temperatures, erratic rainfall, and ecological loss—not with fear, but with foresight.

# Chapter 5: Climate-Responsive Planning

## 5.1 Planning in a Warming World

Climate change is no longer a distant threat—it is a lived reality. Cities across Pakistan and beyond face rising temperatures, erratic rainfall, urban flooding, and heat stress. Yet planning responses remain fragmented, reactive, and often cosmetic. Climate-responsive planning demands more than green rhetoric—it requires structural reform.

This begins with acknowledging that climate risk is spatial. Vulnerability is not evenly distributed. Informal settlements, low-income neighborhoods, and ecologically fragile zones bear the brunt of climate impacts. Planning must therefore prioritize these areas—not just with emergency relief, but with proactive design, zoning, and investment.

Climate is not a constraint—it is a planning mandate. Every overlay, setback, and zoning code must respond to heat, water, and ecological thresholds. When planners treat climate as a technical add-on, they ignore its spatial urgency and ethical weight. Climate-responsive planning is not about mitigation alone—it is about redesigning systems for resilience, equity, and care. Because when climate is sidelined, planning becomes complicit in future harm.

## 5.2 Overlays and Adaptive Zoning

One of the most powerful tools in climate-responsive planning is the overlay—a spatial layer that modifies base zoning to reflect environmental realities. Heat overlays, flood overlays, and ecological buffers can guide development intensity, setback rules, and infrastructure priorities.

Adaptive zoning allows planners to respond to changing conditions. Instead of fixed categories, it uses data-driven thresholds to adjust land use permissions. For example, a flood-prone zone might restrict basement construction or mandate elevated plinths. A heat overlay might require reflective roofing or tree canopy ratios.

These tools are not just technical—they are ethical. They ensure that planning reflects lived risk, not abstract categories.

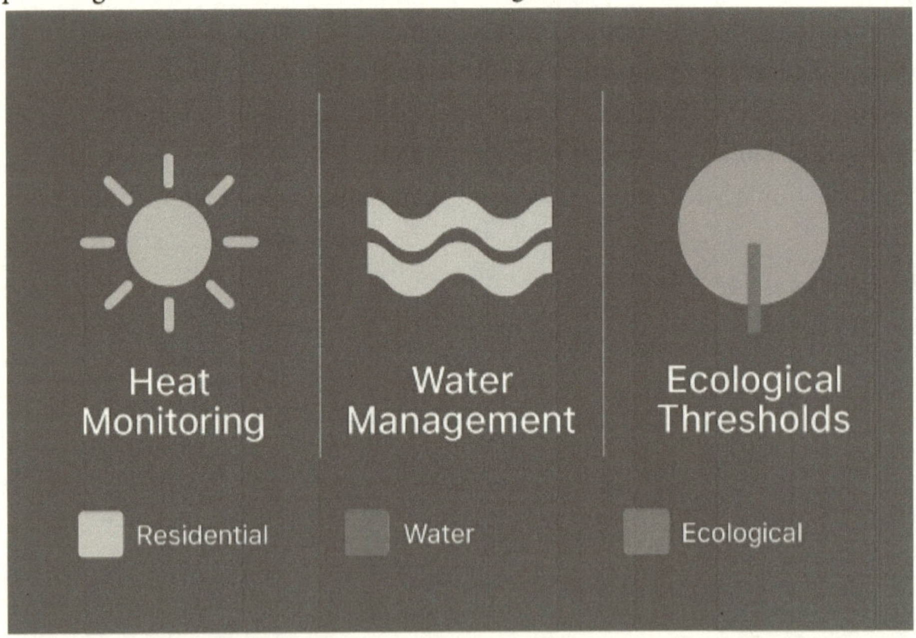


Climate-responsive planning begins with ethical clarity—but it must be translated into spatial tools. Overlays, setbacks, and zoning codes are not just technical instruments; they are civic commitments. When planners embed climate thresholds into land use logic—mapping heat corridors, flood zones, and ecological buffers—they move from intention to infrastructure. The following section explores how adaptive zoning can operationalize this ethic, ensuring that climate is not just acknowledged, but structurally addressed.

## 5.3 Heat-Ready Cities

A recent public reflection explores what it means to build heat-ready cities. In Pakistan, rising temperatures are not just uncomfortable—they are deadly. Urban heat islands, poor ventilation, and asphalt-heavy streets amplify the crisis. Planning must respond with urgency and precision.

Heat-readiness begins with diagnostics: mapping temperature gradients, identifying vulnerable populations, and assessing building materials. It then moves to design: shaded corridors, ventilated housing, reflective surfaces, and green infrastructure. But it must also address governance—ensuring that heat mitigation is embedded in zoning codes, building bylaws, and public procurement.

Heat is not just a climate issue—it is a planning challenge. And planners must lead the response.

# 5.4 Sponge Cities and the Ethics of Absorption

Urban resilience is not just about withstanding shocks—it is about absorbing them. Sponge cities prioritize permeability. They replace concrete surfaces with bioswales, rain gardens, and porous pavements. They integrate wetlands into urban design, restore natural catchments, and reimagine parks as flood buffers.

The ethics of absorption demand that we rethink how cities handle water—not as waste, but as resource. Planning must shift from extraction to retention, from diversion to integration. This means:

- Zoning overlays that mandate permeable surfaces in flood-prone zones
- Design codes that reward green roofs, retention ponds, and tree canopy ratios
- Governance frameworks that align water management with spatial justice

Sponge cities are not just technical models—they are moral propositions. Absorption is not weakness—it is wisdom. It is the planner's ability to listen to terrain, to design with humility, and to align infrastructure with ecological truth.

## 5.5 Terrain-Sensitive Lodging

In ecologically fragile hilly regions, accommodation choices are not merely architectural—they are ecological decisions. A recent public reflection compares the environmental, economic, and experiential impacts of 5-star hotels versus elevated eco-lodging in mountain tourism zones.

Luxury hotels offer comfort and scale but often require extensive land modification, high resource consumption, and centralized ownership. In contrast, elevated eco-lodging—such as tree houses and pods—preserves vegetation, supports biodiversity, and enables community-based tourism.

Eco-tourism must be treated as climate infrastructure. Planning tools such as zoning overlays, environmental impact assessments, and resource management frameworks are essential to operationalize this vision. These tools must prioritize:

- Vegetation preservation through minimal footprint design
- Biodiversity support via clustered, low-impact lodging
- Community benefit through decentralized ownership and local stewardship

This reflection exemplifies how visual storytelling and public discourse can shape ethical planning. It positions the planner not only as a technician, but as a facilitator of spatial justice.

## 5.6 Eco-Tourism as Climate Infrastructure

Eco-tourism is not a peripheral concern—it is a climate-responsive planning strategy. In Pakistan's mountainous and ecologically fragile zones, tourism development must align with terrain sensitivity, community benefit, and ecological integrity.

This means rejecting extractive models—those that flatten forests for resorts or privatize access to heritage—and embracing regenerative frameworks:

- Elevated lodging typologies such as pods and tree houses
- Mobility systems that prioritize electric shuttles and pedestrian trails
- Revenue models that reinvest in local stewardship and biodiversity protection

Eco-tourism overlays must be embedded in zoning codes, supported by environmental impact assessments, and guided by community consultation. These are not optional add-ons—they are planning imperatives.

Eco-tourism is a test of planning integrity. It reveals whether we can align development with ecological ethics, whether we can design for experience without erasing context, and whether we can build futures that are both livable and luminous.

## 5.7 Climate Justice and Planning Ethics

Climate-responsive planning is not just about technical adaptation—it is about justice. Who gets protected? Who gets relocated? Who gets consulted? These are ethical questions that must shape every climate decision.

Planning must center the most vulnerable. It must engage communities not just as data points, but as co-creators. It must recognize that climate resilience is not just physical—it is social, economic, and cultural.

This requires a shift in mindset. From designing for stability to designing for change. From planning for growth to planning for survival. From zoning for control to zoning for care.

Climate justice is not a thematic add-on—it is a planning imperative. When planners center vulnerability, equity, and ecological resilience, they transform zoning and overlays into tools of justice. A visual anchor reinforces this imperative.

## 5.8 Toward a Climate-Responsive Planning Toolkit

To operationalize this vision, planners need modular tools:

- Climate overlays: heat, flood, drought, and ecological buffers
- Responsive zoning codes: adaptive thresholds and differentiated permissions
- Green infrastructure standards: canopy ratios, permeable surfaces, and bioswales
- Community engagement protocols: participatory diagnostics and co-design
- Monitoring dashboards: real-time data for adaptive governance

These tools must be embedded in planning education, consultancy practice, and municipal workflows. They must be accessible, scalable, and context-sensitive.

Climate-responsive planning is not a trend—it is a necessity. And planners must rise to meet it.

## 5.9 Heat as Infrastructure Failure: Planning for Thermal Justice

A recent public reflection reframes heatwaves not as seasonal anomalies, but as systemic urban failures. In Pakistan, rising temperatures expose the fragility of our built environments—especially in low-income settlements where shade, ventilation, and cooling access are scarce.

Heat is not just a climate issue; it is a planning indictment. It must be treated as infrastructure. This calls for:

- Thermal audits to map heat vulnerability
- Chrono-zoning that aligns land use with heat rhythms
- Cooling infrastructure—from tree canopies to public shelters
- Governance reform that embeds heat resilience into zoning codes and budgets

Heatwaves are not evenly distributed. Informal settlements, outdoor laborers, and elderly populations bear the brunt of thermal stress. Yet planning responses remain fragmented and technocratic.

This reflection calls for planners to treat heat as a civic priority—one that demands ethical urgency, spatial intelligence, and operational clarity. Because when heat becomes a hazard, planning must become protection.

Climate-responsive planning is not just about overlays and thresholds—it is about life. The next chapter explores urban wellbeing: how health, dignity, and spatial care must shape the ethics of city-making. Because a breathable city is not enough—it must also be livable.

# Chapter 6: Planning for Urban Wellbeing

Urban wellbeing is not a byproduct of development—it is a planning imperative. In Pakistan and beyond, cities shape health outcomes through air quality, mobility, housing, and access to care. Yet planning frameworks often treat health as peripheral, delegating it to medical systems rather than embedding it in spatial design. This chapter argues that healthy cities must be ethically planned—not just medically managed. It explores how zoning, infrastructure, and governance can either support or undermine wellbeing, and how planners must reclaim health as a civic responsibility.

# 6.1 Planning for Wellbeing: Beyond Infrastructure

Urban wellbeing is not a sectoral concern—it is a spatial imperative. In cities across Pakistan and beyond, the health of urban populations is shaped not only by hospitals and clinics, but by zoning codes, mobility networks, and environmental design. Planning decisions determine exposure to pollutants, access to green space, and the physical and mental resilience of communities. Yet too often, wellbeing is treated as an afterthought—addressed through reactive measures rather than proactive design.

This chapter reframes wellbeing as a planning mandate. It argues that city health frameworks must be embedded in the very logic of urban development: in how we allocate land, regulate density, and design public space. It calls for planners to move beyond technical compliance and embrace ethical responsibility—to treat health not as a deliverable, but as a condition of justice. Urban wellbeing is not a passive result of infrastructure—it is a deliberate planning outcome. Cities shape how we breathe, move, eat, and feel. When planners center mental health, nutrition, sensory experience, and preventive design, they move beyond technical delivery toward civic care. Planning for wellbeing means designing cities that heal, not harm—spaces that support physical vitality, emotional resilience, and public trust.

Planning for urban wellbeing begins with a shift in intent—from infrastructure delivery to civic care. It demands that planners move beyond roads and buildings to consider how cities shape emotion, cognition, and health. When mental health, nutrition, and sensory experience are treated as design inputs—not afterthoughts—planning becomes a tool for healing. The following section explores neuro-urbanism as a framework for designing cities that support emotional resilience and public trust.

# 6.2 Neuro-Urbanism and the Design of Mental Health

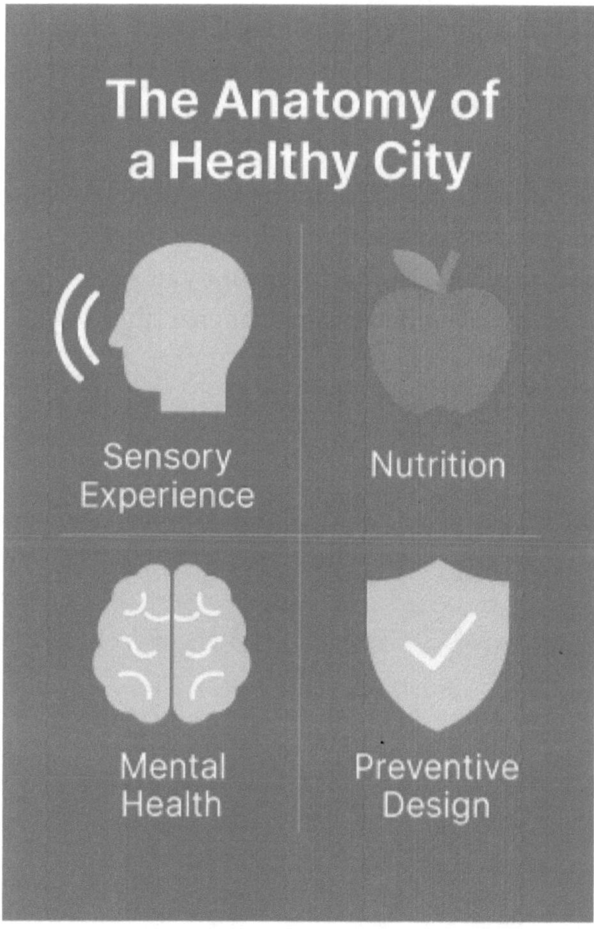


Urban planning has long focused on physical infrastructure—roads, buildings, and zoning codes. But a recent public reflection I shared calls for a deeper lens: *neuro-urbanism*, the study of how urban environments shape mental health. In Pakistan's dense, overstimulated cities, this lens is not academic—it is urgent.

Neuro-urbanism asks: How do noise, congestion, and spatial fragmentation affect cognition, anxiety, and emotional resilience? It challenges planners to design not just for movement, but for mental clarity. This means:

- Quiet zones embedded in master plans.

- Visual rhythm in streetscapes to reduce cognitive fatigue.
- Nature access as a baseline, not a luxury.

Mental health is spatial. A poorly ventilated apartment, a chaotic intersection, or a sterile plaza can trigger stress responses. Conversely, shaded corridors, textured facades, and sensory diversity can promote calm and connection. Planning must recognize these dynamics—not as aesthetics, but as public health infrastructure.

This reflection argues that mental wellbeing must be embedded in zoning codes, design guidelines, and procurement standards. It calls for planners to collaborate with neuroscientists, psychologists, and community advocates. And it positions neuro-urbanism not as a niche, but as a foundational ethic: to build cities that heal, not harm.

## 6.3 Urban Green Spaces — Vital for Health and Activity

Urban green spaces are not decorative—they are ecological infrastructure. They regulate temperature, absorb pollutants, support biodiversity, and offer refuge from the sensory overload of dense cities. In Pakistan's rapidly densifying urban environments, green equity is often overlooked, leaving marginalized communities with little access to shade, biodiversity, or safe outdoor activity.

Green spaces are vital for physical health, mental well-being, and civic engagement. Parks, tree corridors, and wetlands are not amenities—they are rights. Their absence deepens spatial inequality and erodes urban resilience.

Ethical planning demands:

- **Green equity audits** to identify underserved zones and prioritize investment.
- **Zoning overlays** that protect green corridors, ecological buffers, and community gardens.
- **Design strategies** that integrate shade, biodiversity, and accessibility—especially for children, elders, and informal workers.

In Pakistan, where informal settlements and peri-urban zones often host the last remnants of urban nature, planners must restore—not erase—these ecologies. Green spaces must be planned as public health infrastructure, not as ornamental afterthoughts. Because when greenery disappears, so does the city's capacity to breathe, heal, and connect.

## 6.4 Dietetics and the Spatial Logic of Nutrition

Urban planning rarely intersects with dietetics—but it must. A recent public reflection I shared explores how spatial design influences nutritional access, dietary habits, and food-related health outcomes. In Pakistan's rapidly urbanizing regions, the built environment often promotes convenience over nutrition, leading to systemic health risks.

Planning decisions shape what people eat. Fast food corridors, vending zones, and car-centric layouts encourage processed consumption. Meanwhile, fresh produce markets, walkable access to groceries, and community kitchens remain underprioritized. This is not just a public health issue—it is a spatial justice concern.

Dietetics must inform zoning overlays, procurement policies, and mobility design. This means:

- Food access overlays that prioritize fresh produce within walking distance.
- Vendor zoning that limits ultra-processed food clustering near schools and hospitals.
- Design incentives for rooftop gardens, edible landscapes, and community-supported agriculture.

This reflection argues that nutrition is not just a behavioral choice—it is a spatial condition. Planners must collaborate with dietitians, public health experts, and community advocates to embed nutritional wellbeing into urban form. The goal is not just to feed cities, but to nourish them.

## 6.5 Urban Math and the Ethics of Quantification

A recent public reflection I shared challenges the way numbers are used in urban planning—not as tools of insight, but as instruments of abstraction. Metrics dominate our reports: population densities, traffic volumes, housing deficits. Yet behind every number is a lived reality. Urban math must be reframed not as a technical exercise, but as an ethical one.

This reflection argues that planners must interrogate the logic of quantification. What is being counted? What is being excluded? Who defines the thresholds, and who bears the consequences?

Ethical quantification demands:

- Contextual framing: Metrics must be interpreted through terrain, governance, and lived experience
- Narrative clarity: Data must be explained in accessible language, not buried in technical jargon
- Civic alignment: Indicators must reflect public priorities—not just institutional convenience

Urban math is not neutral. It shapes funding, zoning, and public perception. When planners fail to question the numbers, they risk reinforcing inequity. When they engage critically, they transform data into dialogue.

Planning is not just about what we measure—it is about how we interpret, communicate, and act. Urban math must serve the city, not obscure it.

## 6.6 The Spatial Anatomy of Health

Healthy cities are not built through medical infrastructure alone. They emerge from walkable networks, breathable housing, and inclusive public realms. The anatomy of urban health includes:

- Mobility systems that reduce vehicular dependency and promote active transport.
- Green infrastructure that mitigates heat, absorbs pollutants, and supports mental wellbeing.
- Zoning overlays that prioritize access to clinics, parks, and clean air—especially in underserved neighborhoods.

These elements must be planned not in isolation, but as interdependent systems. A shaded corridor is not just a design feature—it is a respiratory intervention. A pedestrian plaza is not just an amenity—it is a mental health asset. The ethics of planning demand that we recognize these connections and act accordingly.

Preventive design is not a luxury—it is civic infrastructure. When planners embed walkability, green buffers, and sensory clarity into urban form, they reduce illness before it begins. Health is not just a sector—it is a spatial ethic. Cities must be designed to support vitality, not just treat disease. Planning for wellbeing means anticipating harm and designing for care.

## 6.7 Sensory Urbanism and the Ethics of City Scents

A recent public reflection I shared explores the role of smell in shaping urban experience. From the jasmine of a courtyard to the diesel of a bus stop, scent is not incidental—it is spatial memory. Yet planning rarely accounts for it.

This reflection argues that sensory urbanism must be treated as a design ethic. Smell mapping is not just poetic—it is diagnostic. It reveals zones of neglect, resilience, and cultural identity.

Ethical planning demands:

- Smell audits alongside noise and air quality assessments
- Preservation of olfactory heritage—from street food corridors to floral zones
- Design strategies that reduce toxic exposure and enhance sensory dignity

Cities are not just seen and heard—they are smelled. When planning ignores this, it erases part of what makes urban life meaningful. Sensory urbanism restores that meaning, one scent at a time.

## 6.8 More Data, Better Planning? The Dangerous Assumption

A recent public reflection I shared critiques a common planning mantra: "more data means better decisions." While data is essential, its uncritical use can be misleading—even harmful.

This reflection argues that data must be contextualized, interpreted, and ethically framed. Raw numbers do not speak for themselves. They reflect who collects them, what gets measured, and what remains invisible.

Ethical planning demands:

- Interrogating data sources—who benefits, who is excluded, and what assumptions shape the metrics
- Balancing quantitative and qualitative insights—fieldwork, lived experience, and civic memory must complement dashboards
- Resisting data absolutism—not every planning challenge can be solved by algorithms or predictive models

Planning is not a spreadsheet—it is a civic practice. When data becomes dogma, we risk designing cities that are efficient but unjust. The goal is not more data. It is better judgment.

## 6.9 Psychogeography and the Emotional Life of Cities

A recent public reflection I shared explores psychogeography—the study of how urban environments shape emotional experience. It asks: how do streets feel? How do corners remember? How does movement become memory?

This reflection argues that planning must account for emotional geographies. Cities are not just built—they are felt. And when planning ignores this, it risks designing spaces that are efficient but alienating.

Ethical planning demands:

- Mapping emotional hotspots—places of joy, grief, protest, and belonging
- Designing for affect—textures, transitions, and thresholds that evoke comfort or curiosity
- Recognizing spatial trauma—sites of displacement, violence, or erasure must be acknowledged and healed

Psychogeography is not a luxury—it is a civic necessity. Because cities are not just navigated. They are lived.

## 6.10 Preventive Design as Public Health

Urban wellbeing must be measured not by emergency response times, but by preventive design. This includes:

- Thermal comfort through reflective surfaces, tree canopies, and ventilated housing.
- Air quality buffers through zoning that separates industrial activity from residential life.
- Mental health infrastructure through access to nature, cultural space, and social cohesion.

Preventive design is not a luxury—it is a necessity. In cities where smog seasons, heatwaves, and noise pollution are normalized, the baseline for livability has eroded. Planning must restore that baseline—not through slogans, but through spatial reform.

Wellbeing is not just a clinical metric—it is a spatial ethic. But health cannot thrive in ecological absence. The next chapter explores how urban ecology must be treated as infrastructure: not as ornament, but as operational resilience.

# Chapter 7: Urban Ecology as Infrastructure

This chapter builds on the civic provocations introduced in Chapter 2, Vanishing Ecologies: Civic Loss and Environmental Injustice. Where Chapter 2 named the sensory and ecological erosion of urban life—smog, heat, noise, and light poverty—this chapter translates those losses into planning responses.

Here, urban ecology is treated not as scenery, but as infrastructure. Each module offers design strategies, zoning tools, and governance frameworks to restore resilience, equity, and ecological integrity. Readers are encouraged to cross-reference both chapters for a full spectrum of insight—from ethical urgency to operational clarity.

## 7.1 The Disappearance of Urban Nature

Urban expansion often comes at the cost of ecological erasure. Wetlands are drained, trees are felled, and biodiversity is displaced to make way for roads, housing, and commercial zones. These losses are rarely accounted for in planning metrics. Environmental impact assessments may be conducted, but mitigation is often superficial. The result is a city that grows in size but shrinks in ecological depth.

Planning must confront this reality. It must recognize that ecological systems are not obstacles—they are assets. They regulate temperature, manage water, support biodiversity, and offer cultural meaning. Their disappearance is not just an environmental concern—it is a planning failure.

Urban ecology is not aesthetic garnish—it is infrastructure. Trees, wetlands, corridors, and deserts are not passive backdrops; they are active systems of resilience, memory, and care. When planners treat ecology as central—not peripheral—they redesign cities to absorb shocks, restore balance, and honor terrain. Ecological planning is not about greening the margins—it is about rethinking the core.

## 7.2 Pollution and the Politics of Visibility

Smog, noise, and waste are not just byproducts of urban life—they are indicators of systemic neglect. In many cities, pollution is normalized. Air quality monitors are absent, waste management is outsourced, and noise regulations are unenforced. These conditions disproportionately affect low-income communities, who live closer to industrial zones, busy roads, and unregulated construction.

Planning must make pollution visible. It must map exposure, enforce accountability, and prioritize remediation. It must treat pollution not as a technical issue, but as a justice issue.

# 7.3 Lost Skies: Light Pollution and the Erasure of Wonder

A reflection I shared publicly explores the quiet tragedy of **light pollution**—how urban brightness erases the night sky, and with it, a sense of cosmic wonder. In many cities, stars are no longer visible. The Milky Way has vanished. Children grow up without ever seeing a constellation.

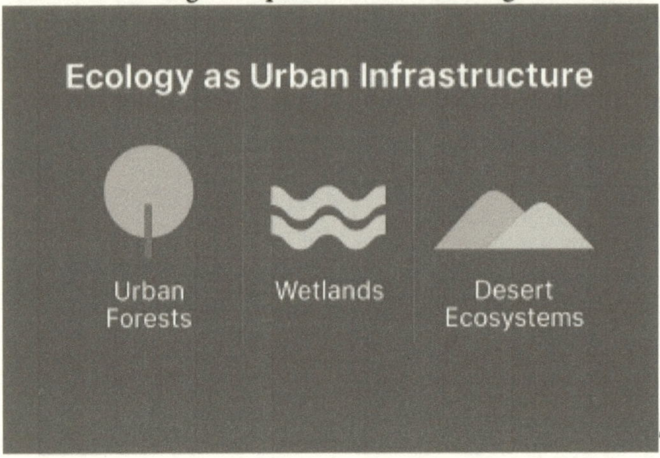


This loss is not just poetic—it is ecological and cultural. Nocturnal species are disoriented. Circadian rhythms are disrupted. Cultural traditions tied to the stars are forgotten. Planning must respond—not by turning off the lights, but by designing with darkness in mind.

This means shielding streetlights, regulating signage, and creating dark-sky corridors. It means recognizing that visibility is not always a virtue, and that some forms of illumination are forms of erasure. Light pollution is a planning issue. And planners must reclaim the night.

## 7.4 Ecological Misalignment: Spatial Fragmentation in Lahore

A recent public reflection I shared critiques Lahore's urban expansion as a geography of fragmentation. Roads are built without transit logic, housing sprawls without service integration, and civic institutions are scattered across disconnected zones. This is not merely poor design—it is ecological misalignment.

This reflection argues that spatial mismatch undermines resilience. It emerges from:

- **Uncoordinated mandates** between planning, transport, and housing authorities
- **Reactive approvals** that prioritize land value over civic coherence
- **Absence of spatial audits** to align infrastructure with population needs

In Lahore, schools are placed far from residential clusters, hospitals are inaccessible by public transit, and green spaces are unevenly distributed. The result is a city that moves inefficiently, governs unevenly, and excludes silently.

Ethical planning demands:

- **Geographic coherence**—where land use, mobility, and services are planned as a system
- **Spatial justice audits**—to identify who benefits and who is left behind
- **Governance integration**—so that planning decisions reflect civic logic, not just bureaucratic convenience

This reflection positions spatial fragmentation as an ecological failure. It calls for planners to treat geography not as backdrop, but as infrastructure. Because when space is misaligned, resilience is compromised—and justice is misplaced.

## 7.5 Pakistan's Deserts — Insights and Development Ethics

A recent public reflection I shared explores Pakistan's deserts—not as barren voids, but as ecological and cultural terrains. From Thar to Cholistan, these landscapes hold memory, resilience, and untapped potential. Yet planning often treats them as blank canvases for extractive development.

This reflection argues that desert planning must be terrain-sensitive. It must respect climatic thresholds, cultural rhythms, and ecological fragility. Deserts are not empty—they are storied.

Ethical planning demands:

- **Climatic audits** before intervention—mapping heat stress, water scarcity, and seasonal rhythms
- **Cultural overlays** that protect nomadic routes, oral geographies, and desert architecture
- **Development strategies** that prioritize adaptation over transformation—solar corridors, water harvesting, and low-impact tourism

Pakistan's deserts are not wastelands. They are living ecologies. Planning must engage them with humility, precision, and care.

# 7.6 Urban Forest Corridors: Ecological Infrastructure for Lahore

A recent public reflection I shared explores the potential of urban forest corridors to transform Lahore's ecological resilience. These corridors are not decorative—they are infrastructure. They regulate temperature, absorb pollutants, support biodiversity, and reconnect fragmented landscapes.

This reflection argues that forest corridors must be treated as planning assets. They offer:

- **Climate mitigation**—reducing urban heat islands and improving air quality
- **Hydrological benefits**—enhancing groundwater recharge and flood absorption
- **Biodiversity support**—creating habitats and migration pathways for urban wildlife
- **Civic engagement**—offering shaded walkways, educational trails, and community stewardship zones

In Lahore, where rapid urbanization has fragmented green space, forest corridors can stitch together parks, canals, and institutional lands into a coherent ecological network. But this requires more than planting trees—it demands zoning overlays, land use reform, and long-term governance.

Ethical planning demands:

- **Ecological mapping**—identifying potential corridors based on terrain, canopy, and connectivity
- **Design standards**—ensuring species diversity, soil health, and climate adaptability
- **Policy integration**—embedding forest corridors into master plans, transport networks, and resilience strategies

This reflection positions urban forestry not as beautification, but as infrastructure. Because when cities breathe through their trees, they become livable, resilient, and just.

## 7.7 Designing the Future City: A Vision for Lahore

A recent public reflection I shared proposes a future-ready vision for Lahore—one that treats ecology, dignity, and spatial coherence as planning infrastructure. This is not a utopian sketch; it is a modular blueprint rooted in terrain, governance, and civic rhythm.

This reflection argues that Lahore's future must be designed through:

- **Ecological integration**—urban forests, canal corridors, and green buffers as climate infrastructure
- **Civic dignity**—public toilets, shaded walkways, and inclusive mobility as baseline services
- **Spatial coherence**—zoning that aligns housing, transit, and services across scales

The vision rejects cosmetic urbanism. It calls for planners to move beyond beautification and embrace ecological logic. In Lahore, where heatwaves, smog, and fragmentation threaten livability, resilience must be designed—not declared.

Ethical planning demands:

- **Modular zoning overlays**—that adapt to climate, density, and civic use
- **Resilience audits**—to assess ecological performance, not just aesthetic appeal
- **Governance alignment**—so that infrastructure reflects public need, not political expedience

This reflection positions Lahore not as a problem to be managed, but as a city to be reimagined. Because the future is not a distant horizon—it is a design decision.

## 7.8 Upgrading vs. Building Anew: Ecological Ethics in Urban Investment

A recent public reflection I shared challenges the default bias toward new construction in urban planning. In Pakistan and beyond, planners often equate progress with new roads, new housing, new flyovers. But this reflection argues that **upgrading existing infrastructure is often more ethical, ecological, and operationally sound**.

The decision to upgrade or build anew is not merely technical—it is ecological. Each choice carries implications for land use, carbon footprint, and community continuity. A visual anchor reinforces this ethical tension.

New development consumes land, displaces communities, and strains ecological systems. It often prioritizes visibility over viability. In contrast, upgrading:

- **Preserves civic memory**—retaining the spatial logic and social rhythms of existing neighborhoods
- **Reduces ecological footprint**—minimizing land disturbance, material use, and carbon emissions
- **Improves equity**—by reinvesting in underserved areas rather than expanding affluent enclaves

This reflection calls for planners to treat **upgrading as a design discipline**, not a compromise. It demands:

- **Audit frameworks**—to assess when upgrading is more resilient than rebuilding
- **Policy reform**—to incentivize adaptive reuse, retrofitting, and ecological restoration
- **Narrative shift**—so that civic pride is tied to stewardship, not spectacle

Ethical planning is not about building more—it is about building wisely. Because resilience is not measured by what is new, but by what endures.

## 7.9 Sanitation as Ecological Infrastructure: The Public Toilet Crisis

A recent public reflection I shared highlights a neglected but essential dimension of urban ecology: public toilets. In Pakistan's cities, the absence of safe, accessible, and dignified sanitation facilities is not merely a logistical oversight—it is an ecological and civic failure.

This reflection argues that sanitation must be treated as infrastructure. It affects:

- **Gender equity**—women, elders, and caregivers face disproportionate barriers to mobility and participation
- **Public health**—lack of facilities contributes to disease, contamination, and informal waste disposal
- **Urban resilience**—sanitation access shapes how cities absorb footfall, host events, and support inclusive public life

In Lahore and other cities, public toilets are rare, poorly maintained, and often unsafe. Their absence restricts access to parks, markets, and transit hubs. It forces vulnerable populations into undignified choices and undermines the promise of inclusive urbanism.

Ethical planning demands:

- **Sanitation audits**—mapping where facilities exist, where they are needed, and who is excluded
- **Design standards**—ensuring toilets are safe, gender-inclusive, and accessible to all
- **Ecological integration**—embedding sanitation into zoning codes, green infrastructure, and civic design frameworks

This reflection calls for planners to treat sanitation not as a peripheral service, but as ecological infrastructure. Because when toilets vanish, so does dignity. And when dignity is compromised, resilience is impossible.

## 7.10 Ecological Memory and Urban Design

Cities must remember what they have erased. This means integrating ecological memory into design—restoring native vegetation, commemorating lost landscapes, and embedding biodiversity into public space. It means designing parks that reflect local ecosystems, not imported aesthetics. It means treating every tree, stream, and bird as part of the city's heritage.

Ecological memory is not nostalgia—it is resilience. It helps cities adapt to climate stress, reconnect with place, and rebuild relationships between people and nature.

Urban terrain is not blank—it is layered with memory, ecology, and meaning. Deserts, wetlands, and forest corridors are not obstacles to development—they are archives of resilience. When planners erase terrain, they erase history. Ecological planning begins by listening to land—its thresholds, its rhythms, its stories. Because cities that forget their terrain cannot design for resilience.

## 7.11 Toward Ecological Planning

To operationalize ecological planning, practitioners need tools that go beyond compliance:

- **Biodiversity overlays**: mapping species corridors and habitat zones
- **Green infrastructure codes**: mandating permeable surfaces, native plants, and ecological buffers
- **Pollution dashboards**: tracking air, noise, and waste exposure
- **Dark-sky zoning**: regulating light intensity and direction
- **Ecological audits**: assessing planning decisions for biodiversity impact

These tools must be embedded in planning education, consultancy practice, and municipal workflows. They must treat ecology not as a constraint, but as a core design principle.

Vanishing ecologies are not inevitable. They are the result of choices. And planners must choose differently.

Urban ecology is not limited to biodiversity, water, or air—it includes memory, ritual, and cultural rhythm. As planners embrace ecological audits and biodiversity overlays, they must also account for intangible infrastructure: the spatial traditions, informal economies, and civic anchors that shape how cities breathe socially. The next module turns to one such anchor—the bazaar—not as a relic, but as a living ecology of commerce, connection, and cultural continuity.

## 7.12 Bazaars vs Malls — Preserving Local Culture

A recent public reflection I shared explores the cultural tension between traditional bazaars and modern malls. From Anarkali to Emporium, the shift is not just architectural—it is ideological. Bazaars are porous, adaptive, and rooted in local rhythms. Malls are enclosed, curated, and driven by consumption logic.

This reflection argues that urban planning must recognize bazaars as cultural infrastructure. They are not obsolete—they are living archives of civic memory, economic resilience, and spatial democracy.

Ethical planning demands:

- Preservation of historic marketplaces through adaptive reuse, pedestrianization, and cultural zoning
- Recognition of informal economies as legitimate contributors to urban vitality
- Design strategies that prioritize openness, accessibility, and local identity over privatized enclosure

Malls may offer comfort, but bazaars offer connection. Planning must not erase the latter in pursuit of the former. It must balance modernization with memory, and development with dignity.

## 7.13 Urbanisation Is Inevitable—Not Culture Erasure

A recent public reflection I shared challenges the assumption that urbanisation must erase local culture. Development is often framed as a binary: modernity or memory, infrastructure or identity. But this framing is false—and dangerous.

This reflection argues that urbanisation must be culturally literate. It must preserve spatial rituals, architectural vernaculars, and community rhythms. Cities are not blank slates—they are layered histories.

Ethical planning demands:

- **Cultural audits** before redevelopment—mapping intangible heritage, spatial customs, and civic rituals
- **Design strategies** that integrate local materials, forms, and narratives
- **Policy safeguards** that protect cultural anchors—shrines, courtyards, bazaars, and oral geographies

Urbanisation is inevitable. But culture erasure is not. Planning must evolve without amnesia. Because cities are not just built—they are remembered.

Urban ecology is not just green—it is grounded. But grounding must extend beyond parks and overlays to the terrains themselves. The next chapter explores fragile landscapes: how mountains, coasts, and deserts demand terrain-sensitive tourism, and how planning must learn to listen before it builds.

# Chapter 8: Designing for Fragile Terrain

# 8.1 Planning in Fragile Terrain

Mountainous regions, coastal belts, and desert margins present unique planning challenges. Their topography, ecology, and cultural heritage demand sensitivity—not standardization. Yet many planning interventions treat these landscapes as blank canvases, imposing grids, roads, and resorts without regard for slope, soil, or spirit.

Fragile terrain is not just physically delicate—it is culturally and ecologically rich. Planning must begin with humility. It must ask: What does this land want? What stories does it hold? What risks does it carry?

Fragile terrain is not a constraint to be overcome—it is a responsibility to be honored. Mountains, deltas, deserts, and wetlands carry ecological thresholds that cannot be ignored. When planners treat vulnerability as a design input—not a nuisance—they protect both landscape and livelihood.

Designing for fragile terrain means listening to land, respecting its limits, and building with care—not conquest.

Planning in fragile terrain demands more than restraint—it demands typological care. When landscapes are vulnerable, every built form must respond to slope, ecology, and memory. Lodging is not just shelter—it is a spatial signal. It tells the land whether it is being respected or consumed. The following section explores how terrain-sensitive lodging can honor ecological thresholds while supporting human presence.

# 8.2 Terrain-Sensitive Lodging

A recent public reflection I shared explores the ethics of **terrain-sensitive lodging.** In Pakistan's northern regions, tourism has surged—but so has ecological damage. Hotels are built on unstable slopes, sewage flows into rivers, and architectural styles erase local identity.

This reflection calls for a new approach: one that respects slope, soil, and settlement patterns. Terrain-sensitive lodging means designing with the land, not against it. It means using local materials, minimizing excavation, and aligning built form with natural contours.

It also means regulating density. A fragile valley cannot host unlimited tourists. Carrying capacity must guide zoning, investment, and infrastructure. Planning must protect the landscape—not just accommodate demand.

## 8.3 Slope-Sensitive Design and Zoning

Slope is not just a technical parameter—it is a planning ethic. Steep slopes require special treatment: limited construction, reinforced foundations, and runoff management. Yet many zoning codes ignore slope altogether, applying flatland standards to vertical terrain.

Slope-sensitive zoning must be embedded in planning practice. It must guide plot sizes, building heights, and access routes. It must integrate geotechnical data, hydrological models, and ecological overlays. And it must be enforced—not just drafted.

This is not just about safety—it is about stewardship. Slope-sensitive design protects lives, landscapes, and legacies.

## 8.4 Eco-Tourism and Planning Ethics

Eco-tourism is often marketed as sustainable—but without ethical planning, it becomes extractive. Resorts may claim green credentials while displacing communities, degrading ecosystems, and commodifying culture.

This reflection argues that eco-tourism must be rooted in planning ethics. It must prioritize local benefit, ecological integrity, and cultural respect. It must engage communities as co-owners, not just service providers. And it must be regulated—not just romanticized.

Planning tools for ethical eco-tourism include:

- **Zoning overlays** for ecological buffers and cultural heritage zones
- **Density caps** based on carrying capacity
- **Design codes** for vernacular architecture and low-impact infrastructure
- **Community protocols** for benefit-sharing and grievance redressal

Eco-tourism is not just a business model—it is a planning responsibility.

## 8.5 Toward Terrain-Sensitive Planning

To operationalize terrain-sensitive planning, practitioners need modular tools:

- **Slope overlays**: guiding construction intensity and access
- **Fragility indices**: mapping ecological and cultural sensitivity
- **Vernacular design codes**: promoting local materials and forms
- **Tourism zoning dashboards**: tracking density, impact, and compliance
- **Community engagement frameworks**: ensuring participatory stewardship

These tools must be embedded in planning education, consultancy practice, and municipal workflows. They must treat terrain not as a constraint, but as a collaborator.

Fragile landscapes are not obstacles—they are teachers. And planners must learn to listen.

## 8.6 Civic Infrastructure in Tourist Landscapes

Tourist zones are not exempt from civic responsibility. Roads, sanitation, water supply, and public space must be planned with both visitors and residents in mind. Yet in many fragile regions, civic infrastructure is reactive—built to serve peak seasons, not long-term resilience.

Planning must treat civic infrastructure as terrain-sensitive. This means:

- **Integrated mobility**—walkable trails, low-impact transit, and safe access for locals
- **Waste management overlays**—regulating disposal, recycling, and seasonal surges
- **Water stewardship**—protecting aquifers, regulating extraction, and promoting reuse
- **Public space design**—ensuring plazas, viewpoints, and rest areas serve both tourists and communities

Tourism must not displace civic logic. It must reinforce it. Because when infrastructure is seasonal, resilience is compromised—and dignity is deferred.

## 8.7 Governance and Stewardship in Fragile Zones

Planning in fragile landscapes requires more than technical expertise—it demands governance reform. Many tourist zones suffer from fragmented mandates, seasonal enforcement, and extractive investment. Stewardship must be institutionalized.

Ethical governance demands:

- **Multi-agency coordination**—aligning tourism, environment, and municipal departments
- **Seasonal zoning codes**—adapting density and access based on ecological thresholds
- **Community stewardship boards**—empowering locals to co-manage tourism flows
- **Monitoring dashboards**—tracking ecological impact, civic performance, and visitor behavior

Tourism is not just a design challenge—it is a governance test. And fragile zones deserve resilient institutions.

# Chapter 9: Mobility and Urban Access

## 9.1 Mobility as a Planning Imperative

Mobility is not just about movement—it is about access, opportunity, and dignity. Cities that fail to provide equitable mobility entrench exclusion. Planning must treat mobility as a core infrastructure of justice, not a secondary concern.

This means designing for all modes: walking, cycling, public transit, and private vehicles. It means prioritizing connectivity, safety, and affordability. And it means recognizing that mobility is not just technical—it is social.

Mobility is not just about moving—it is about reaching. It is about who gets to participate in urban life, and who is left waiting. Roads, transit lines, and walkways are not neutral—they encode equity, rhythm, and opportunity. Ethical mobility planning begins by asking: who is excluded, and why? Because when access is uneven, movement becomes privilege—not infrastructure.

Mobility is not just a technical challenge—it is a civic reckoning. When planners treat movement as infrastructure, they risk ignoring who gets to move, and how. But when mobility is reframed as access, every transit corridor becomes a question of equity. The following section explores how Transit-Oriented Development (TOD) and modal equity can help planners design cities that move with their people—not just around them.

# 9.2 Transit-Oriented Development (TOD) and Modal Equity

A recent public reflection I shared explores the ethics of **Transit-Oriented Development (TOD)**. TOD is often celebrated for its efficiency and density—but without equity, it becomes exclusionary. High-end TOD zones may offer sleek transit access while pricing out low-income residents.

This reflection calls for **modal equity**—ensuring that all forms of movement are supported, and that TOD serves diverse populations. It urges planners to integrate affordable housing, pedestrian infrastructure, and inclusive zoning into TOD frameworks.

TOD must be more than a design strategy—it must be a justice strategy. TOD is not just about transit—it is about trust. A recent design reflection explores how Lahore's evolving transit corridors must be framed not only by infrastructure, but by equity. Modal equity demands that TOD serve all users—not just the mobile elite. Planners must align transit investment with civic inclusion, spatial justice, and governance reform.

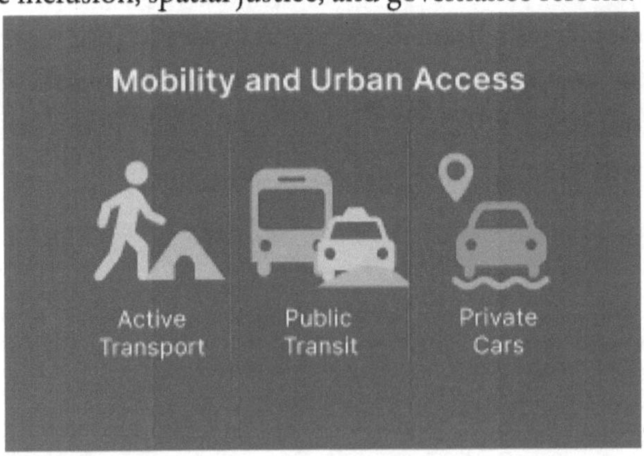


TOD is not just about transit—it is about trust. A recent design reflection explores how Lahore's evolving transit corridors must be framed not only by infrastructure, but by equity. Modal equity demands that TOD serve all users—not just the mobile elite. Planners must align transit investment with civic inclusion, spatial justice, and governance reform.

TOD is also not just a spatial strategy—it is a civic ethic. A recent design reflection explores how Lahore's transit corridors must be reframed through the lens of access, affordability, and governance. Modal equity demands that TOD serve all users—not just the mobile elite. Planners must interrogate land use, fare structures, and institutional mandates to ensure that transit investment delivers justice—not just infrastructure.

## 9.3 Walkability and Urban Rhythm

Walkability is often reduced to sidewalk width and street furniture. But true walkability is about rhythm, safety, and sensory experience. It is about how people feel when they move through space—whether they are welcomed, protected, and inspired.

Planning for walkability means designing continuous paths, shaded corridors, and active edges. It means minimizing vehicular dominance and maximizing human scale. And it means listening to walkers—not just measuring footfall.

# 9.4 Reimagining Service Areas Along M2 Motorway

A recent public reflection I shared critiques the design logic of motorway service areas—spaces meant for rest, refueling, and refreshment. Yet many are sterile, generic, and disconnected from local context. They serve vehicles, not people.

This reflection argues that service areas must be reimagined as civic infrastructure. They are not just logistical nodes—they are spatial experiences. Their design shapes how travelers feel, interact, and remember the journey.

Ethical planning demands:

- **Contextual design**—using local materials, vernacular forms, and regional identity
- **Inclusive amenities**—clean restrooms, shaded seating, prayer spaces, and safe zones for women and children
- **Ecological integration**—native landscaping, water harvesting, and low-carbon construction

Motorways are arteries—but service areas are their pulse. Planning must treat them not as afterthoughts, but as civic invitations. Because every stop is a chance to restore dignity, memory, and care.

## 9.5 Mobility Dashboards and Planning Tools

To operationalize mobility planning, practitioners need modular tools:

- **Modal overlays**: mapping pedestrian, cycling, and transit zones
- **TOD zoning codes**: integrating density, access, and affordability
- **Walkability audits**: assessing safety, continuity, and comfort
- **Mobility dashboards**: tracking usage, gaps, and equity metrics

Mobility is not just about getting from A to B—it is about who gets to move, how, and why.

## 9.6 Car-Centric Cities and the Ethics of Urban Design

A recent public reflection I shared challenges the dominance of car-centric planning in Pakistani cities. Roads widen while sidewalks shrink. Flyovers multiply while pedestrian crossings vanish. This is not just a design flaw—it is a moral failure.

Car-centric cities prioritize speed over safety, asphalt over access, and private mobility over public wellbeing. They marginalize pedestrians, cyclists, and transit users—especially women, children, and the elderly. Planning must confront this imbalance.

Ethical urban design must:

- Reclaim public space for people, not vehicles
- Embed walkability and modal equity into master plans
- Prioritize human scale in zoning, streetscape, and mobility audits
- Challenge the logic of flyovers and expressways that fragment communities and erase civic rhythm

Car-centric planning is not neutral—it is exclusionary. It reflects a value system that privileges speed, consumption, and privatization. Ethical planning must reverse this logic. It must design cities that welcome, not repel. That connect, not divide.

## 9.7 Street Vendors and Pedestrian Rights

A recent public reflection I shared challenges the invisibility of pedestrians and street vendors in urban mobility discourse. Sidewalks are narrowed, crossings erased, and vending zones criminalized—all in the name of traffic flow. But cities are not highways. They are habitats of encounter, exchange, and movement.

This reflection argues that pedestrian rights must be reframed as civic rights. Walking is not a residual mode—it is the most democratic form of mobility. And street vendors are not obstructions—they are economic actors, cultural anchors, and spatial stewards.

Ethical mobility planning demands:

- **Sidewalk dignity**: Design must prioritize width, shade, safety, and accessibility
- **Vendor inclusion**: Planning must accommodate informal economies through designated zones, flexible permits, and spatial audits
- **Pedestrian rhythm**: Crossings, buffers, and public furniture must reflect how people actually move—not just how vehicles flow

Mobility is not just about transport—it is about belonging. When pedestrians are treated as obstacles and vendors as nuisances, the city loses its civic soul. Planning must restore that soul by designing for presence, not just passage.

## 9.8 Mobility Dashboards and Planning Tools

To operationalize mobility planning, practitioners need modular tools:

- **Modal overlays**: mapping pedestrian, cycling, and transit zones
- **TOD zoning codes**: integrating density, access, and affordability
- **Walkability audits**: assessing safety, continuity, and comfort
- **Mobility dashboards**: tracking usage, gaps, and equity metrics

Mobility is not just about getting from A to B—it is about who gets to move, how, and why.

A recent public reflection I shared challenges the dominance of car-centric planning in Pakistani cities. Roads widen while sidewalks shrink. Flyovers multiply while pedestrian crossings vanish. This is not just a design flaw—it is a moral failure.

Car-centric cities prioritize speed over safety, asphalt over access, and private mobility over public wellbeing. They marginalize pedestrians, cyclists, and transit users—especially women, children, and the elderly. Planning must confront this imbalance.

Ethical urban design must:

- Reclaim public space for people, not vehicles
- Embed walkability and modal equity into master plans
- Prioritize human scale in zoning, streetscape, and mobility audits
- Challenge the logic of flyovers and expressways that fragment communities and erase civic rhythm

Car-centric planning is not neutral—it is exclusionary. It reflects a value system that privileges speed, consumption, and privatization. Ethical planning must reverse this logic. It must design cities that welcome, not repel. That connect, not divide.

## 9.9 Cycling as Ecological Infrastructure

A recent public reflection IZHAR UL HAQ shared reframes cycling not as a lifestyle choice, but as ecological infrastructure. In Pakistan's congested cities—where smog, noise, and fuel dependency dominate the urban experience—cycling offers a low-emission, high-impact alternative. Yet it remains marginal, underserved by policy, infrastructure, and civic imagination.

This reflection argues that cycling must be embedded in planning frameworks—not as recreation, but as resilience. Ethical urbanism demands:

- **Dedicated cycling corridors** that are safe, shaded, and accessible across income groups
- **Zoning overlays** that prioritize non-motorized mobility in dense and vulnerable zones
- **Civic campaigns** that normalize cycling as a dignified, equitable mode of transport

Cycling is not just about movement—it is about air quality, spatial justice, and urban rhythm. When cities invest in cycling, they invest in breathable futures.

Cycling is also not just mobility—it is ecological infrastructure. A recent design reflection explores how cycling networks in Pakistan can serve climate resilience, modal equity, and spatial justice. From shaded corridors to safe intersections, cycling must be planned as a civic right—not a recreational privilege. Planners must embed cycling into zoning, overlays, and investment logic—not just pilot projects.

# Chapter 10: Spatial Justice and the Right to the City

## 10.1 What Is Spatial Justice?

Spatial justice is the ethical distribution of space, access, and opportunity. It asks: Who gets to live where? Who has access to parks, schools, and transit? Who is displaced, and who is protected?

Planning decisions shape these outcomes. Zoning codes, infrastructure investments, and land use policies determine who benefits and who bears the cost. Spatial justice demands that these decisions be made transparently, inclusively, and equitably.

Inclusion is not a policy add-on—it is a spatial practice. When cities are designed with marginalized voices at the center, justice becomes visible in form, function, and access. A visual anchor reinforces this imperative.

## 10.2 Exclusion by Design

Many cities are designed to exclude. Gated communities, exclusionary zoning, and privatized public spaces reinforce social hierarchies. Informal settlements are criminalized, while luxury enclaves are subsidized. Planning must confront these patterns—not perpetuate them.

Exclusion is not accidental—it is architectural. And justice must be built into the blueprint. Public space is not neutral—it is an ethical terrain. When benches deter rest, fences restrict movement, or lighting patterns signal exclusion, design becomes a tool of control. A visual anchor reinforces this spatial critique.

Spatial justice is not a slogan—it is a reckoning. It asks who is allowed to belong, and who is designed out. Every gate, overlay, and zoning code carries a politics of access. The right to the city is not just about being present—it is about being visible, valued, and safe. When planners confront exclusion by design, they begin the work of justice—not as outcome, but as process.

Exclusion is not accidental—it is designed. When planners map access, they also map denial. Gated enclaves, fragmented transit, and privatized public space are not just urban features—they are spatial verdicts. But justice begins when these verdicts are challenged. The right to the city is not a slogan—it is a civic entitlement. It demands that every resident be seen, heard, and included—not just in policy, but in place.

## 10.3 The Right to the City

A reflection I shared publicly revisits the concept of **the right to the city**—not as a slogan, but as a planning ethic. This right includes access to housing, mobility, public space, and participation. It is not granted—it is claimed.

This reflection argues that planners must act as stewards of this right. They must design cities that welcome, not repel. They must challenge exclusionary codes, amplify marginalized voices, and democratize urban space.

The right to the city is not abstract—it is spatial, political, and urgent.

## 10.4 Toward Spatial Justice Planning

To embed spatial justice in planning practice, practitioners need tools that reveal and repair inequality:

- **Equity overlays**: mapping access to services, amenities, and infrastructure
- **Displacement dashboards**: tracking eviction, gentrification, and tenure loss
- **Inclusive zoning codes**: mandating affordability, accessibility, and diversity
- **Participatory planning protocols**: ensuring community voice and veto

Spatial justice is not a chapter—it is a lens. And every planning decision must pass through it.

## 10.5 Land as Memory, Land as Power

Land is not just a commodity—it is memory, identity, and power. Yet planning often treats land as a blank slate, ignoring histories of displacement, dispossession, and resistance.

Spatial justice demands that land governance be reimagined. This means:

- **Tenure audits**—mapping ownership, occupation, and eviction patterns
- **Restorative zoning**—prioritizing housing and services in historically excluded areas
- **Land value capture**—redirecting speculative gains toward public benefit
- **Cultural overlays**—protecting sacred sites, ancestral commons, and informal heritage

Land is not neutral. It remembers. And planning must learn to listen.

Land is not neutral—it is narrative. A recent design reflection explores how land politics shape identity, exclusion, and resistance in urban Pakistan. From informal settlements to elite enclosures, the struggle over land is a struggle over belonging. Planners must treat land not as empty space, but as lived history—where memory, power, and justice collide.

## 10.6 Justice as Process, Not Just Outcome

Spatial justice is not only about where parks are built or who gets housing—it is about who decides, who speaks, and who is heard.

Justice is procedural. It requires:

- **Transparent planning workflows**—open data, accessible hearings, and clear appeals
- **Community veto rights**—especially in projects affecting tenure, culture, or ecology
- **Equity impact assessments**—evaluating who benefits and who bears the cost
- **Planning literacy campaigns**—enabling communities to understand and shape urban decisions

When planning is opaque, justice is deferred. When it is participatory, justice becomes possible.

# 10.7 Visualizing Spatial Justice

Spatial justice is not an abstract principle—it is a visible condition. The figure below distills three recurring patterns in exclusionary urban design and civic response. Each icon represents a spatial logic that planners must interrogate, reform, and reimagine.

• Gated Enclaves

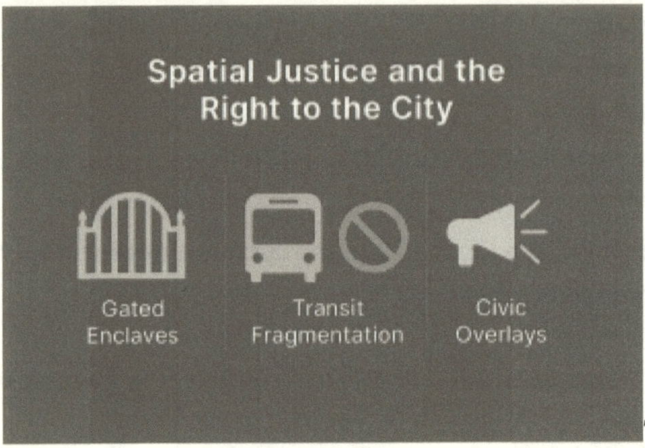


These are not just physical barriers—they are spatial verdicts. Gated communities privatize access, fragment civic rhythm, and encode exclusion. Their design logic often reflects fear, privilege, and withdrawal from shared urban life.

• Transit Fragmentation

When transit systems are disconnected, modal equity collapses. Fragmented routes, uneven coverage, and inaccessible stops disproportionately affect low-income and marginalized communities. Transit must be planned as civic infrastructure—not as a patchwork of convenience.

• Privatized Public Spaces

Parks, plazas, and walkways are increasingly fenced, monetized, or surveilled. This erodes the right to linger, gather, and belong. Public space must remain public—not just in name, but in experience. Spatial justice demands visibility, safety, and civic entitlement.

These icons are not just illustrations—they are planning prompts. They ask: who is excluded, and how? And they remind us that justice must be designed—not just declared.

# Chapter 11: Visual Planning and Modular Clarity

## 11.1 Why Visual Planning Matters

Urban planning is a visual discipline. Maps, overlays, diagrams, and dashboards are not just illustrations—they are arguments. They shape how problems are framed, how solutions are proposed, and how decisions are justified. Yet many planning documents suffer from visual clutter, technical opacity, or editorial inconsistency.

Visual clarity is not cosmetic—it is operational. It determines whether stakeholders understand, engage, and act. Poor visuals lead to misalignment, delays, and mistrust. Planning must therefore treat visual design as a core competency—not a post-production task.

Visual planning is not about beautification—it is about comprehension. Diagrams, inserts, and modular layouts are not aesthetic flourishes; they are planning tools. When visuals clarify complexity, they become civic infrastructure. When formatting guides the reader, it becomes pedagogical ethics. In planning, clarity is not optional—it is a public responsibility.

## 11.2 Modular Formatting as Planning Infrastructure

Modular formatting is more than a stylistic choice—it is a planning infrastructure. It allows documents to be updated, reused, and adapted without losing coherence. It supports teaching, publishing, and consultancy workflows. And it enables planners to track inserts, overlays, and thematic shifts with precision.

Modular clarity includes:

- **Numbered sections** with thematic anchors
- **Tagged inserts** for case studies, reflections, and overlays
- **Visual consistency** across diagrams, captions, and legends
- **Editable surfaces** for future-proofing and stakeholder input

This approach reflects your manuscript's ethos: clarity, adaptability, and operational rigor.

## 11.3 Visual Storytelling in Planning

A reflection I shared publicly explores the power of **visual storytelling** in planning. Technical brilliance means little if stakeholders cannot grasp the narrative. A well-designed map can reveal injustice. A clear diagram can unlock consensus. A dashboard can guide real-time decisions.

This reflection argues that planners must become visual strategists. They must learn to layer information, highlight priorities, and simplify complexity. They must design for comprehension—not just compliance. Visual storytelling is not decoration—it is civic communication.

Visual planning is not decorative—it is pedagogical. A recent video walkthrough demonstrates how design logic can be translated into modular clarity. The sequence illustrates how ethical planning principles—accessibility, operational alignment, and civic relevance—can be embedded directly into visual formats. When design carries rationale, it becomes a planning tool, not just a presentation layer.

## 11.4 AI-Powered Design Tools for Planners

Planners need tools that support clarity without requiring graphic design expertise. AI-powered platforms now offer:

- **Auto-layout dashboards** for zoning, mobility, and overlays
- **Smart captioning** that adapts to audience literacy
- **Template libraries** for modular inserts and visual glossaries
- **Real-time collaboration** for stakeholder review and annotation

These tools democratize visual planning. They allow consultants, officials, and educators to produce high-impact visuals with minimal friction. And they align with your goal of operational clarity across formats.

## 11.5 Toward a Visual Planning Toolkit

To embed visual clarity in planning practice, practitioners need modular tools:

- **Overlay templates**: for zoning, climate, mobility, and equity
- **Insert trackers**: for modular reflections and case studies
- **Visual glossaries**: for demystifying technical language
- **Formatting protocols**: for section numbering, captioning, and layout
- **Design dashboards**: for real-time editing and stakeholder input

These tools must be embedded in planning education, consultancy practice, and publishing workflows. They must treat visual clarity not as a luxury, but as a standard.

Planning is not just about what we build—it is about how we communicate. And visual clarity is the foundation of trust.

**11.6 Visual Ethics and Civic Trust**

Visuals are powerful—but power must be ethical. A map can reveal injustice, but it can also conceal it. A diagram can clarify complexity, but it can also distort it. Planning must confront the ethics of visual representation.

Visual exclusion is subtle. Informal settlements may be left off maps. Pedestrian zones may be rendered invisible. Cultural sites may be reduced to icons. These omissions are not technical—they are political.

Ethical visual planning demands:

- **Transparency in data sourcing and layering**—so that stakeholders know what is shown, and what is not
- **Inclusive representation**—of informal economies, pedestrian rhythms, and cultural memory
- **Accessible design**—for diverse literacy levels, visual impairments, and civic audiences
- **Narrative integrity**—so that visuals reflect lived experience, not just technical abstraction

Visual clarity is not just about aesthetics—it is about trust. And trust must be designed.

# Chapter 12: Affordable Housing and the Ethics of Shelter

## 12.1 Housing as a Human Right

Housing is not just a commodity—it is a human right. Yet in many cities, planning systems treat housing as a market product, prioritizing profit over dignity. Affordable housing is sidelined, informal settlements are criminalized, and public housing is underfunded. Planning must reclaim housing as a civic obligation.

This means recognizing that shelter is foundational. It shapes health, education, mobility, and opportunity. It determines who belongs in the city—and who is pushed out. Shelter is not a commodity—it is a civic ethic. When planners prioritize dignity, tenure security, and spatial inclusion, housing becomes a platform for justice.

## 12.2 The Affordable Housing Crisis in Punjab

A recent public reflection I shared confronts the deepening crisis of affordable housing in Punjab. Despite policy rhetoric and pilot schemes, the reality on the ground remains stark: land prices soar, informal settlements expand, and middle-income families are priced out of dignity.

This reflection argues that housing is not just a commodity—it is a civic right. Ethical planning demands:

- **Land audits** to identify underutilized public land for affordable housing
- **Zoning reforms** that enable mixed-income development and prevent speculative exclusion
- **Design strategies** that prioritize dignity, climate resilience, and community integration

In Punjab, the housing crisis is not a failure of construction—it is a failure of governance, imagination, and equity. Planning must restore housing as a public good, not a private windfall.

Affordability is not just about cost—it is about access, dignity, and governance. A recent design reflection explores how public-private joint ventures are reshaping the housing landscape in Punjab. These partnerships, when ethically framed, can bridge the gap between policy ambition and delivery reality. But planners must interrogate the terms—who benefits, who is excluded, and what civic values are embedded in the deal.

Housing is not just about walls and roofs—it is about dignity, safety, and belonging. When affordability is reduced to market logic, entire communities are priced out of existence. Ethical planning demands that shelter be treated as civic infrastructure—not as speculative real estate. Because when housing fails, cities fracture. And when housing is just, cities heal.

## 12.3 The Politics of Affordability

Affordability is not just about price—it is about access, tenure, and stability. A unit may be "affordable" on paper but inaccessible due to location, discrimination, or hidden costs. Planning must interrogate affordability metrics and ensure they reflect lived realities.

This includes:

- **Location-sensitive affordability**: proximity to jobs, schools, and transit
- **Tenure diversity**: rental, ownership, cooperative, and transitional models
- **Anti-displacement safeguards**: zoning protections and rent stabilization

Affordability is not a threshold—it is a continuum. And planning must respond accordingly.

Affordability is not just a market failure—it is a planning failure. When housing is treated as a commodity, exclusion becomes systemic. But planners are not powerless. They can reframe shelter as civic infrastructure, redesign zoning for equity, and anchor housing within growth pole strategies. The next section explores how spatial planning can restore affordability—not through subsidy alone, but through structural alignment.

## 12.4 Growth Poles and Housing Equity

A recent reflection I shared explores the concept of **growth poles**—urban nodes that attract investment, infrastructure, and migration. While growth poles can drive development, they often exacerbate housing inequality. Land prices surge, informal settlements expand, and low-income residents are pushed to peripheries.

This reflection argues that planners must embed housing equity into growth pole strategies. This means reserving land for affordable housing, integrating mixed-income zoning, and ensuring that infrastructure benefits all residents—not just investors. Growth must be inclusive—or it is not growth at all.

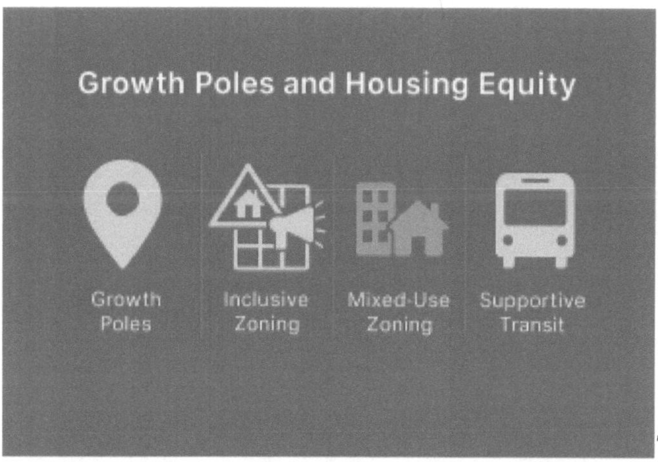


The accompanying visual distills four critical planning tools that shape the ethics of shelter and the spatial logic of affordability. The first icon, a teardrop-shaped map marker, represents Growth Poles—strategic anchors for housing investment and infrastructure alignment. These poles decentralize affordability and enable planners to distribute shelter access beyond elite urban cores. The second icon, a zoning grid overlaid with a megaphone, symbolizes Inclusive Zoning—a civic tool that amplifies community voice and enables mixed-income housing within regulatory frameworks. The third icon pairs a low-rise commercial block with a pitched-roof house, illustrating Mixed-Use Zoning as a strategy to integrate residential, commercial, and civic functions, thereby enhancing walkability and land-use efficiency. The final icon, a front-facing bus, signifies Supportive Transit, reinforcing the principle that housing equity must be linked to mobility access. Together, these icons offer a modular planning toolkit—one that reframes shelter not as a commodity, but as civic infrastructure.

# 12.5 Urban Housing in Pakistan — Beyond Supply and Demand

A recent public reflection I shared critiques the dominant framing of housing in Pakistan. It is often reduced to a supply-demand equation, ignoring the deeper civic, spatial, and ethical dimensions. Housing is not just shelter—it is belonging, stability, and dignity.

This reflection argues that housing must be treated as civic infrastructure. It is not a product—it is a right. And when planning treats it as a commodity, it reinforces exclusion, speculation, and spatial injustice.

**Ethical housing policy demands:**

- **Decommodification of shelter**—recognizing housing as a public good, not just a private asset
- **Inclusive zoning and tenure reform**—to protect renters, informal dwellers, and low-income households
- **Design strategies for dignity**—ventilation, daylight, privacy, and community integration must be non-negotiable

Pakistan's housing crisis is not just about numbers—it is about narratives. We must shift from building units to building futures. Because housing is not just where we live—it is how we are seen.

## 12.6 Informal Settlements and Planning Ethics

Informal settlements are often framed as problems to be solved. But they are also solutions—self-built responses to housing exclusion. Planning must shift from eradication to integration. It must recognize the ingenuity, resilience, and community embedded in informal housing.

This means:

- **In-situ upgrading**: improving infrastructure without displacement
- **Tenure regularization**: granting legal recognition and protection
- **Participatory design**: involving residents in planning and implementation

Informality is not failure—it is adaptation. And planning must learn from it.

## 12.7 Toward a Shelter-Responsive Planning Toolkit

To embed housing ethics in planning practice, practitioners need modular tools:

- **Affordability overlays**: mapping price, access, and tenure diversity
- **Growth pole zoning codes**: integrating equity into development nodes
- **Informality dashboards**: tracking tenure, infrastructure, and risk
- **Housing audits**: assessing availability, quality, and displacement risk
- **Community protocols**: co-designing housing upgrades and safeguards

Shelter is not just a roof—it is a right. And planning must protect it.

## 12.8 Housing as Climate Resilience

Housing is not just about walls—it is about weather. In floodplains, heat islands, and seismic zones, shelter must do more than protect—it must adapt. Climate vulnerability begins at home. And planning must respond.

Ethical housing policy demands:

- **Climate overlays**—guiding housing placement, density, and materials based on ecological risk
- **Resilient design codes**—ensuring ventilation, drainage, shading, and thermal comfort
- **Disaster-safe zoning**—protecting vulnerable communities from floods, landslides, and heat stress
- **Green retrofitting protocols**—upgrading existing housing stock for resilience and efficiency

Climate resilience is not a luxury—it is a necessity. And every home must be a haven, not a hazard.

Affordable housing is not just a neighborhood issue—it is a regional imperative. When shelter is planned in isolation, it risks reinforcing urban cores while neglecting peripheral communities. But when housing is embedded within regional planning frameworks, it becomes a tool for spatial equity, climate resilience, and inclusive growth. The next chapter expands this lens, exploring how planners can align shelter, mobility, and infrastructure across scales—moving from site-level interventions to regional strategies that serve entire populations.

# Chapter 13: Regional Planning and the Ethics of Scale

## 13.1 Why Regional Planning Matters

Cities do not exist in isolation. They are part of regional systems—linked by rivers, roads, economies, and ecologies. Yet planning often stops at municipal boundaries, ignoring the interdependencies that shape urban life. Regional planning is not just a technical necessity—it is an ethical imperative.

It allows planners to coordinate infrastructure, manage migration, balance growth, and protect shared resources. It enables cities to act not as competitors, but as collaborators. And it ensures that development is distributed—not concentrated.

Pakistan's urban future is unfolding rapidly—marked by demographic shifts, infrastructural strain, and governance complexity. A recent design reflection outlines key opportunities and challenges, reinforcing the urgency of regional planning. From peri-urban expansion to climate corridors, planners must engage with scale not as abstraction, but as operational reality. Regional planning is not a luxury—it is a civic imperative.

Urban transformation is not a distant goal—it is unfolding in real time. A recent design reflection explores how Punjab's cities are navigating governance reform, spatial expansion, and civic urgency. These shifts demand more than infrastructure—they require ethical regional planning. Planners must align scale with stewardship, ensuring that transformation serves equity, not just efficiency.

Regional planning is not about drawing bigger maps—it is about thinking with coherence, equity, and care. Every road, corridor, and growth pole affects lives beyond its immediate footprint. Scale-sensitive planning demands that we coordinate across jurisdictions, align infrastructure with terrain, and distribute opportunity—not just concentrate it. Because when planning ignores scale, it reinforces fragmentation. And when scale is treated ethically, it becomes a tool of justice.

## 13.2 Growth Poles and Regional Equity

A reflection I shared earlier explores the concept of **growth poles**—urban nodes that attract investment, infrastructure, and migration. While growth poles can drive development, they often create regional imbalances. Smaller towns are neglected, rural areas are drained, and infrastructure is concentrated in select corridors.

This reflection argues that planners must embed **regional equity** into growth pole strategies. This means:

- Linking growth poles to surrounding towns through transit and investment
- Reserving land for affordable housing and public services
- Ensuring that infrastructure benefits are shared—not hoarded

Growth must be distributed—or it becomes distortion.

## 13.2.1 Urban Density and the Ethics of Vertical Growth

Urban density is often celebrated as a marker of progress—especially in cities like Lahore, where vertical expansion is equated with modernization. But density is not inherently ethical, equitable, or sustainable. It must be planned with care, contextualized within ecological thresholds, and aligned with community needs.

This reflection challenges the assumption that verticality equals smartness. In Lahore's high-rise corridors, speculative development often outpaces infrastructure, governance, and public oversight. The result is a landscape of exclusion: gated towers with private amenities, disconnected from the city's broader mobility, affordability, and resilience goals.

Smartness must be redefined. It is not about sensors or skylines—it is about ecological intelligence, inclusive zoning, and terrain-sensitive planning. In cities where basic services remain uneven, smartness must prioritize access, not aesthetics.

Planning for density requires more than technical tools. It demands ethical judgment, participatory engagement, and strategic restraint. Fast-track approvals, unregulated verticality, and imported design templates undermine long-term resilience. Planners must slow down, think clearly, and act with integrity.

## 13.3 Strategic Recommendations:

- **Redefine smartness**: Prioritize ecological and social intelligence over digital infrastructure.
- **Audit vertical growth**: Use zoning overlays and environmental impact assessments to regulate high-rise corridors.
- **Promote density with dignity**: Ensure dense neighborhoods have access to green space, clean air, and public services.
- **Engage communities**: Use participatory mapping and citizen assemblies to shape density decisions.

## 13.4 Regional Planning — Equity Beyond the City

A recent public reflection I shared calls for a shift in urban discourse: from city-centric planning to *regional equity*. In Pakistan, urban development often focuses on metropolitan cores—Lahore, Karachi, Islamabad—while surrounding districts remain under-resourced, disconnected, and institutionally neglected. This imbalance deepens spatial injustice and undermines national resilience.

This reflection argues that regional planning is not just logistical—it is ethical. It determines who gets infrastructure, who accesses opportunity, and who is left behind. When ring roads, economic corridors, and housing schemes are planned without regional coordination, they reinforce fragmentation. When planning is siloed by jurisdiction, it fails to reflect how people actually live, move, and work.

Ethical regional planning demands:

- **Inter-district coordination**—aligning transport, housing, and service delivery across administrative boundaries
- **Equity audits**—mapping resource distribution, access gaps, and institutional capacity regionally
- **Regional governance frameworks**—empowering planning bodies with mandate clarity and budgetary autonomy
- **Data integration**—using shared platforms to track mobility, climate risk, and economic flows across regions

In Pakistan, where migration, climate stress, and economic interdependence blur city boundaries, planning must evolve. A metro line in Lahore affects housing in Sheikhupura. A flood in Dera Ghazi Khan disrupts supply chains in Multan. Regional planning is not optional—it is foundational.

This reflection calls for planners to think beyond municipal limits. To treat regions as ecosystems. And to design with alignment, not isolation.

The tension between efficiency and equity is not just technical—it is ethical. Regional planning must move beyond urban-centric models and embrace distributive justice. The video reflection underscores this imperative, challenging planners to design systems that serve peripheries, not just power centers.

City growth is not random—it is coded. A recent design reflection decodes how urban expansion in Punjab reflects governance gaps, spatial improvisation, and civic urgency. Growth must be understood not just as physical sprawl, but as a planning challenge. Planners must ask: What logic drives expansion? Who benefits from it? And how can regional planning restore equity across the urban-rural continuum?

Regional planning is not just about scale—it is about coherence. In Pakistan, urban expansion often reflects governance fragmentation, not strategic foresight. A recent design reflection explores how Lahore's growth patterns reveal deeper institutional gaps, spatial improvisation, and civic exclusion.

City growth is not neutral—it encodes governance. A recent design reflection decodes how Lahore's expansion reveals fragmented mandates, unregulated sprawl, and exclusionary planning logic. From peri-urban ribbon development to elite enclaves, the city's growth trajectory demands a regional lens. Planners must ask: What logic drives this expansion? Who is left behind? And how can regional planning restore equity across the urban-rural continuum?

## 13.5 Regional Planning in Developing Contexts — Beyond the Blueprint

A recent public reflection I shared outlines the persistent challenges of planning in developing countries—particularly in Pakistan. These challenges are not confined to cities; they are regional in nature. Fragmented mandates, weak enforcement, and disconnected planning documents reflect systemic governance failures across jurisdictions.

This reflection argues that regional planning must be reframed as a governance reform strategy. In many districts, master plans are produced but rarely implemented. Zoning regulations exist but are inconsistently enforced. Public participation is invited but rarely empowered. The result is a planning culture that is performative rather than transformative.

Ethical regional planning demands:

- **Mandate clarity** across jurisdictions
- **Implementation audits** that track delivery, not just documentation
- **Capacity building** for regional institutions and civic literacy
- **Political insulation** to protect planning from short-term interference
- **Community accountability** embedded in regional feedback loops

In Pakistan, where urban growth is rapid and uneven, regional planning must evolve from static blueprints to adaptive systems. It must respond to terrain, governance, and public trust. This reflection calls for planners to act not just as technicians, but as reformers—bridging the gap between policy and practice and restoring integrity to the regional planning process.

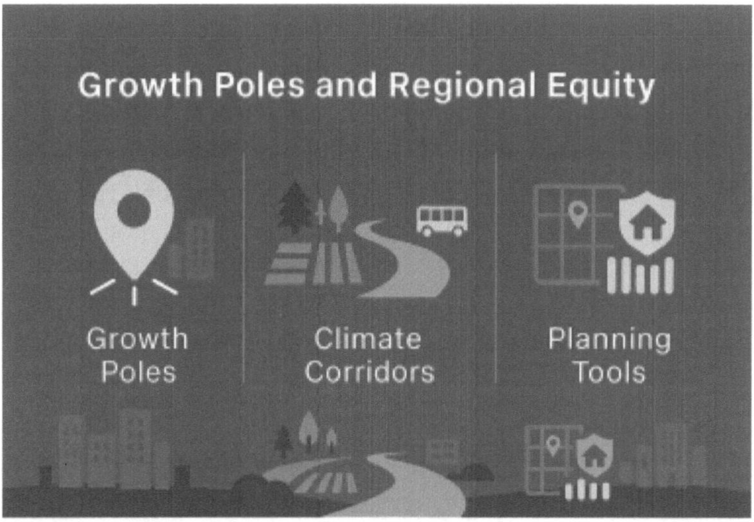


Regional planning in developing contexts cannot rely on idealized blueprints. A recent design reflection explores how corruption, instability, and injustice distort urban futures—turning plans into performances and infrastructure into inequity. Planners must confront these realities with operational clarity and civic courage. Ethics are not an overlay—they are the foundation.

The visual accompanying this chapter illustrates three foundational tools for ethical regional planning. The first icon, labeled Growth Poles, features a central orange map pin connected by white transit arms to adjacent urban clusters. This symbolizes decentralized hubs that anchor infrastructure, housing, and economic opportunity across regions—moving beyond core-centric development. The second icon, titled Climate Corridors, depicts a winding river flanked by forests, farmland, and a transit route. This composition conveys the integration of ecological buffers with sustainable mobility, reinforcing resilience and continuity across terrain. The third icon, under Planning Tools, combines a geographic grid with a location pin, a housing shield, and ascending bar graphs. Together, these elements represent the operational instruments planners must deploy to coordinate across jurisdictions, monitor equity outcomes, and align infrastructure with both topography and civic need. Each icon is rendered in a muted urban palette—slate blue, soft grey, and warm orange—ensuring visual clarity and thematic resonance.

Greenfield development is not a blank canvas—it is a contested terrain. A recent design reflection explores how urban expansion into undeveloped land demands more than technical feasibility. It requires ethical framing, ecological foresight, and civic accountability. In developing contexts, planners must interrogate the logic of growth: Who defines it? Who benefits? And what is lost in the process?

## 13.6 Scale-Sensitive Planning

Planning at scale requires a shift in mindset. It means thinking beyond plot sizes and zoning codes to regional flows, ecological corridors, and economic gradients. It means recognizing that decisions made in one city affect others—through migration, pollution, and resource use.

Scale-sensitive planning includes:

- **Inter-city coordination**: shared transit, housing, and climate strategies
- **Ecological regionalism**: managing watersheds, forests, and biodiversity across boundaries
- **Economic balancing**: supporting secondary cities and rural hubs

Scale is not just a measurement—it is a responsibility.

## 13.7 Regional Planning Tools and Protocols

To operationalize regional planning, practitioners need modular tools:

- **Growth pole overlays**: mapping investment nodes and equity corridors
- **Regional dashboards**: tracking migration, infrastructure, and ecological stress
- **Transit integration codes**: aligning mobility across jurisdictions
- **Inter-city planning protocols**: enabling shared governance and funding
- **Scale audits**: assessing the ripple effects of urban decisions

Regional planning is not about control—it is about coordination. And planners must lead that shift.

## 13.8 Climate Corridors and Regional Resilience

Climate does not respect boundaries. Floods, droughts, and heatwaves move across districts, disrupting lives, economies, and infrastructure. Regional planning must therefore embed climate corridors—zones of ecological risk, migration pressure, and adaptive infrastructure.

This means:

- **Watershed overlays**—mapping floodplains, aquifers, and drainage networks across jurisdictions
- **Climate migration dashboards**—tracking displacement, tenure loss, and service gaps
- **Resilient infrastructure codes**—aligning roads, housing, and utilities with terrain and risk
- **Inter-district coordination protocols**—ensuring emergency response and resource sharing across boundaries

Climate resilience is not local—it is regional. And planning must scale accordingly.

Regional planning may coordinate across jurisdictions, but it is planning education that shapes the values behind those decisions. As growth poles and climate corridors become tools of equity, the question arises: who is training the next generation to wield them responsibly? The final chapter turns inward—toward the institutions, ethics, and pedagogical frameworks that will define the future of planning practice. It asks not only what planners must know, but how they must be formed.

# Chapter 14: Planning Education and the Future of Practice

Planning instruction often emphasizes technical skills—GIS, zoning codes, infrastructure design. But it must also cultivate ethical judgment, civic imagination, and narrative clarity. Planners are not just technicians—they are storytellers, strategists, and stewards of public trust.

Educational practice must reflect this complexity. It must teach students to read cities critically, engage communities respectfully, and communicate ideas clearly. It must balance rigor with readability, and theory with practice.

## 14.1 Modular Teaching and Adaptive Curriculum

Modular teaching allows educators to adapt content to context. Instead of rigid syllabi, it offers thematic inserts, case studies, and visual tools that can be rearranged, updated, and localized. This approach mirrors your manuscript's architecture—modular, audience-ready, and future-proof.

Modular curriculum supports:

- **Contextual relevance**: adapting inserts to local challenges
- **Instructional flexibility**: teaching ethics, mobility, climate, and zoning as standalone or integrated modules
- **Collaborative learning**: enabling students to co-create overlays, audits, and planning tools

Planning instruction must evolve from static lectures to dynamic, modular engagement.

## 14.2 Mentorship and Ethical Stewardship

A reflection I shared publicly explores the role of **mentorship** in planning education. Technical mastery is important—but ethical stewardship is learned through example, dialogue, and reflection. Mentors shape how planners think, speak, and act.

This reflection argues that educators must model clarity, humility, and accountability. They must challenge students to interrogate assumptions, embrace complexity, and prioritize justice. They must treat teaching not as instruction—but as co-creation.

Mentorship is not a role—it is a responsibility.

## 14.3 Modes of Engagement: Direct vs. Institutional Mediation

In the realm of professional services—whether in consulting, design, legal work, or technical projects—the choice between working directly with clients or through a firm greatly influences the dynamics of the relationship and the quality of outcomes.

Engaging with clients directly allows for immediate feedback and clarification of objectives, leading to fewer misunderstandings and misinterpretations. Clients can express their needs and expectations without filters, enabling service providers to respond more effectively. This direct interaction fosters stronger relationships, as building personal connections enhances trust and collaboration. In these relationships, the nuances of conversation can facilitate deeper mutual understanding, making it easier to navigate challenges that may arise during the project.

Another advantage of direct collaboration is the ability to tailor solutions to meet specific client needs. Direct access to key stakeholders enables service providers to gain a rich understanding of the client's unique challenges and cultural context. This understanding leads to customized, innovative solutions, and providers can adapt to feedback in real time, producing outcomes that align closely with the client's vision. Flexibility is also enhanced in this model. Direct engagement allows for quick pivots in project direction or scope since service providers can react agilely without navigating bureaucratic layers typical of larger firms. Creative brainstorming sessions can occur more freely, leading to innovative problem-solving opportunities that may be stifled in more hierarchical structures.

Faster decision-making is a crucial benefit of working directly with clients. The streamlined communication process accelerates approvals and feedback loops, leading to quicker project progression. However, there are also challenges to consider. Professionals working independently may face resource limitations, lacking access to the extensive networking and institutional support that firms provide.

Working through a firm comes with its own set of challenges. Communication barriers can arise from layers of hierarchy and bureaucracy, which may lead to misunderstandings or delays. Information often travels through multiple channels, potentially diluting the clarity of the communication. The sense of personalization can diminish, making clients feel more like a number than a valued partner, especially when multiple teams handle different aspects of the project.

Despite the merits of both approaches, many professionals argue that working directly with clients often brings about better outcomes. The potential for clearer communication, stronger relationships, customized solutions, increased flexibility, and faster decision-making presents a compelling case for direct engagement.

## 14.4 From Aspirational to Actionable: Breaking the Cycle of Unimplemented Plans

As an urban planner, I often find myself reflecting on a pressing issue that seems to pervade our field: the repeated cycle of planning studies and master plans that ultimately fail to see effective implementation. Despite the diligent effort and hard work that goes into these projects, many remain unrealized or only partially executed. This raises a crucial question: What is fundamentally wrong with our planning processes?

We recognize some of the barriers to implementing these plans, yet we continue to approach the situation in the same manner, expecting different outcomes. Could it be that the frameworks we create lack the necessary structure and adaptability to meet current demands?

Transforming our planning governance and methodologies is undoubtedly a complex challenge, particularly in the short term. However, simply sticking to established concepts and methods will not yield meaningful change. The planning paradigm we employ must evolve to align with our existing systems, or we risk wasted efforts and resources.

Looking back, we can see numerous initiatives like the National Human Settlements Policy Study or the Management of Cities Policy, not to mention the National Reference Manual, which garnered significant acceptance. More recently, we've seen the Punjab Spatial Strategy and many urban master plans that have been developed. Yet, what has become of these efforts? They often remain on paper, gathering dust rather than being translated into tangible improvements for our communities.

It is imperative that we pause, reflect, and engage in a comprehensive assessment of these plans. We need to suggest practical solutions that make them not just aspirational, but actionable and implementable. The time has come for us to innovate our approach to urban planning, ensuring that we create frameworks that are resilient, inclusive, and capable of addressing the diverse needs of our populations.

Let us come together as a community of planners, policymakers, and stakeholders to embark on this journey of transformation. By fostering open dialogue and collaborative efforts, we can develop strategies that not only facilitate the implementation of our plans but also make a lasting impact on the cities and communities we serve.

## 14.5 Contractual Discipline and Ethical Delivery Models

It is imperative that the firm or firms awarded a project refrain from subcontracting the work to smaller companies. This practice poses significant risks, as the primary contractor is ultimately accountable for the quality of the work and adherence to deadlines. If issues arise, such as poor workmanship or delays in project delivery, it is the firm that has been awarded the contract which will bear the brunt of the client's dissatisfaction and potential penalties.

Moreover, clients should exercise caution when disbursing project fees. If a substantial portion of the payment is released upfront, without correspondence to the actual progress of the work completed, the primary contractor may lose motivation and commitment to the project. This can lead to a decline in the quality and timeliness of the work, as the contractor may shift focus away from the project once they have received significant financial compensation.

Therefore, a balanced approach to payment schedules that aligns with milestones achieved is crucial in ensuring that the contractor remains engaged and dedicated to delivering a high-quality outcome on time. Ethical delivery models must be reinforced not only through planning frameworks but also through contractual discipline, financial accountability, and transparent client-contractor relationships.

Contracts may define scope and timelines, but delivery defines trust. Ethical planning is not just about what gets built—it is about how, for whom, and under what terms. When planners treat delivery models as civic commitments, not just technical obligations, they restore integrity to the profession. Discipline without ethics is compliance. Discipline with ethics is stewardship.

## 14.6 Radical Reform for Service Delivery: A Planner's Call to Action

Improved service delivery in urban planning is not merely a technical challenge—it is a systemic imperative. The repeated failure of planning frameworks to translate into tangible outcomes demands a radical shift in how we approach governance, accountability, and implementation.

Too often, planners and policymakers operate within rigid institutional boundaries that stifle innovation and responsiveness. The result is a cycle of aspirational documents, disconnected from the realities of urban life. To break this cycle, we must confront the structural inertia embedded in our planning systems and embrace a transformative mindset.

This transformation begins with acknowledging that our current models are insufficient. We must move beyond incremental adjustments and pursue bold reforms that prioritize transparency, responsiveness, and community-centered delivery. This includes rethinking how projects are scoped, how stakeholders are engaged, and how performance is measured—not just in terms of outputs, but in terms of lived impact.

Planners must lead this shift by advocating for frameworks that are not only technically sound but ethically grounded. We must champion models that empower local actors, streamline decision-making, and ensure that service delivery is not an afterthought but a core design principle.

The call to action is clear: we need a new planning ethos—one that is agile, inclusive, and uncompromising in its commitment to real-world outcomes. Only then can we restore public trust and deliver the urban futures our communities deserve.

## 14.7 Transforming Land Transactions: Toward Transparent and Ethical Systems

Land transactions in Punjab remain one of the most opaque and contested domains in urban development. Despite decades of reform efforts, the process continues to be riddled with inefficiencies, mistrust, and procedural ambiguity. For planners and policymakers committed to ethical practice, this is not just a technical challenge—it is a moral imperative.

The transformation of land transaction systems must begin with a clear-eyed assessment of existing bottlenecks: fragmented records, discretionary approvals, and inconsistent valuation mechanisms. These flaws not only delay development but also erode public confidence in planning institutions.

A transparent, digitized, and verifiable land transaction framework is essential. This includes the integration of GIS-based land registries, automated valuation protocols, and real-time tracking of approvals. Such systems reduce the scope for manipulation and ensure that land-related decisions are based on data, not discretion.

Equally important is the need for public-facing platforms that allow citizens to verify land status, ownership history, and transaction timelines. When communities can access and understand the systems that govern land, they are more likely to trust and participate in planning processes.

Ethical land governance also requires capacity-building within institutions. Training officials to interpret digital records, enforce standardized procedures, and engage transparently with stakeholders is critical to sustaining reform.

Punjab's recent initiatives—such as digitized cadastral mapping and spatial planning pilots—offer a promising foundation. But without a systemic shift in how land transactions are conceptualized, monitored, and enforced, these efforts risk becoming isolated successes.

Planners must advocate for land systems that are not only efficient but also just. Only then can we ensure that land, as a foundational asset of urban development, serves the public good rather than private interests.

Technology is not a substitute for ethics—but it can be a powerful ally. In Punjab, digital platforms are beginning to transform land transactions, reducing opacity and enhancing public trust. A recent design reflection illustrates how digitization can support ethical delivery—streamlining verification, minimizing discretion, and embedding accountability into everyday processes. Planning reform must embrace these tools not as technical upgrades, but as civic infrastructure.

## 14.8 Global Best Practices: Learning from International Planning Models

Urban planning is not a static discipline—it evolves through the exchange of ideas, experiences, and innovations across borders. As planners in Pakistan confront persistent challenges in implementation, governance, and equity, it is imperative to look outward and learn from global best practices.

Cities around the world have pioneered models that balance growth with sustainability, participation with efficiency, and innovation with tradition. From Singapore's integrated land use and transport systems to Copenhagen's climate-responsive urban design, these examples offer more than inspiration—they offer operational templates.

However, the goal is not replication. It is contextual adaptation. Global models must be interpreted through the lens of local realities: institutional capacity, cultural norms, and regulatory frameworks. What works in Barcelona or Seoul may not translate directly to Lahore or Faisalabad—but the principles of transparency, modularity, and stakeholder engagement can be reimagined to suit our context.

Benchmarking against international standards also helps planners advocate for reform. When local stakeholders see what is possible elsewhere, it creates pressure to modernize systems, digitize workflows, and prioritize inclusive outcomes. It also fosters professional pride and a sense of global belonging among planners who often feel isolated within bureaucratic constraints.

Pakistan's planning community must embrace this global dialogue—not as passive observers, but as active contributors. By documenting local innovations, participating in international forums, and integrating best practices into curricula and consultancy models, we can elevate the discipline and deliver better outcomes for our cities.

The future of planning lies in ethical adaptation, not blind imitation. Let us learn globally, act locally, and build systems that reflect both our aspirations and our realities.

Learning from global best practices requires more than replication—it demands contextual adaptation. A recent design reflection explores how international models can inform terrain-sensitive planning in Pakistan. From participatory zoning in Latin America to climate-responsive overlays in Southeast Asia, these examples offer operational insights. But planners must interrogate each model's assumptions, governance logic, and civic relevance before integration. Global learning must be filtered through local ethics.

## 14.9 AI and the Future of Planning Practice

AI is reshaping planning workflows—from zoning dashboards to participatory mapping. But its integration must be ethical, transparent, and inclusive. Planners must learn to use AI not as a shortcut, but as a collaborator.

This includes:

- **AI-powered design tools**: for overlays, audits, and visual storytelling
- **Data ethics protocols**: ensuring privacy, consent, and bias mitigation
- **Collaborative platforms**: enabling real-time stakeholder engagement

AI can enhance planning—but it cannot replace judgment. Instructional frameworks must prepare planners to lead with conscience, not just code.

## 14.10 Toward a Future-Ready Planning Toolkit

To prepare students and practitioners for ethical, adaptive planning, educators need modular tools:

- **Insert libraries**: thematic reflections and case studies
- **Overlay templates**: for zoning, climate, mobility, and equity
- **Audit protocols**: for housing, pollution, and displacement
- **Visual glossaries**: demystifying technical language
- **Mentorship frameworks**: guiding ethical dialogue and reflection

Planning education must be modular, ethical, and future-ready. It must teach not just how to plan—but why, for whom, and with what consequences.

Smart urban futures are not built by sensors alone—they require policy coherence, ethical framing, and civic alignment. A recent design reflection explores how planners must bridge governance gaps, integrate adaptive tools, and strengthen institutional protocols. The future-ready toolkit is not just technical—it is ethical, modular, and context-sensitive.

## 14.11 Planning Literacy and Civic Education

Planning is not just for professionals—it is for publics. Yet many communities remain excluded from planning conversations due to jargon, opacity, and institutional distance. This exclusion is not accidental—it is architectural.

Planning literacy must become a civic priority. It empowers citizens to understand, question, and shape the decisions that govern their cities. It transforms passive recipients into active participants.

This means:

- **Accessible glossaries**—demystifying technical terms, zoning codes, and planning protocols
- **Community workshops**—teaching overlays, audits, and participatory tools in local languages
- **Public dashboards**—visualizing infrastructure, budgets, and service gaps in real time
- **Youth engagement programs**—embedding planning ethics and spatial awareness in school curricula
- **Civic feedback loops**—ensuring that public input is not symbolic, but structural

Planning literacy is not a supplement—it is a safeguard. And every citizen deserves to understand the city they inhabit.

# Chapter 15: Planning Glossary and Modular Index

## 15.1 Why Glossaries Matter

Planning language is often dense, technical, and exclusionary. Terms like "setback," "overlay," or "TOD" may be familiar to professionals but opaque to communities. A glossary is not just a reference—it is a tool of inclusion. It demystifies jargon, clarifies intent, and supports civic understanding.

This glossary reflects commitment to clarity, accessibility, and modular teaching. It is designed for students, consultants, officials, and community stakeholders alike.

## 15.2 Glossary of Planning Terms

| Term | Definition |
|---|---|
| Overlay | A spatial layer that modifies base zoning to reflect environmental, social, or infrastructural realities. |
| Setback | The minimum distance a building must maintain from a property line, road, or other boundary. |
| TOD (Transit-Oriented Development) | A planning strategy that concentrates housing, jobs, and amenities around public transit hubs. |
| Modular Insert | A standalone reflection, case study, or thematic section that can be embedded or taught independently. |
| Growth Pole | A strategic urban node that attracts investment, infrastructure, and migration, often used in regional planning. |
| Slope-Sensitive Zoning | Land use regulation that accounts for terrain steepness, soil stability, and runoff risk. |
| Dark-Sky Corridor | A planning designation aimed at reducing light pollution and preserving night sky visibility. |
| Responsive Zoning | Adaptive land use regulation that adjusts based on environmental data, social needs, or climate risk. |
| Visual Glossary | A diagrammatic or illustrated reference tool that explains planning terms through images and overlays. |
| Visual Ethics | The practice of designing maps, diagrams, and dashboards with transparency, inclusivity, and civic integrity. |
| Planning Literacy | The ability of citizens and students to understand, question, and shape urban planning decisions. |
| Contractual Discipline | Structuring project agreements to ensure milestone-based accountability and ethical delivery. |
| Implementation Audit | A protocol for tracking whether planning frameworks translate into real-world outcomes. |
| Land Governance Reform | Systemic transformation of land transaction systems to ensure transparency, efficiency, and public trust. |
| Benchmarking | Comparing planning models across cities or countries to |

| | |
|---|---|
| | inspire contextual reform. |
| AI-Powered Planning Tools | Platforms that support overlays, audits, and stakeholder engagement using artificial intelligence. |
| Planning Toolkit | A modular set of inserts, overlays, and protocols designed for teaching, consultancy, and governance. |

## 15.3 Modular Insert Index

| Insert ID | Title | Chapter | Section |
|---|---|---|---|
| MI-01 | Terrain-Sensitive Lodging | 7 | 7.2 |
| MI-02 | Planning Theory and the Evolution of Urban Practice | 2 | 2.3 |
| MI-03 | Demystifying Planning Language | 2 | 2.4 |
| MI-04 | Heat-Ready Cities | 5 | 5.3 |
| MI-05 | TOD and Modal Equity | 8 | 8.2 |
| MI-06 | Partnerships as Planning Infrastructure | 3 | 3.3 |
| MI-07 | Zoning and the Ethics of Control | 4 | 4.3 |
| MI-08 | Logarithmic Zoning | 4 | 4.5 |
| MI-09 | Lost Skies: Light Pollution and the Erasure of Wonder | 6 | 6.3 |
| MI-10 | Eco-Tourism and Planning Ethics | 7 | 7.4 |
| MI-11 | The Right to the City | 9 | 9.3 |
| MI-12 | Visual Storytelling in Planning | 10 | 10.3 |
| MI-13 | Growth Poles and Housing/Regional Equity | 12 | 12.2 |
| MI-14 | Mentorship and Ethical Stewardship | 13 | 13.3 |
| MI-15 | Visual Ethics and Civic Trust | 11 | 11.6 |
| MI-16 | Housing as Climate Resilience | 12 | 12.8 |
| MI-17 | Climate Corridors and Regional Resilience | 13 | 13.7 |
| MI-18 | Planning Literacy and Civic Education | 14 | 14.11 |
| MI-19 | Contractual Discipline and Ethical Delivery Models | 14 | 14.5 |
| MI-20 | Radical Reform for Service Delivery | 14 | 14.6 |
| MI-21 | Transforming Land Transactions | 14 | 14.7 |
| MI-22 | Global Best Practices | 14 | 14.8 |
| MI-23 | AI and the Future of Planning Practice | 14 | 14.9 |
| MI-24 | Toward a Future-Ready Planning Toolkit | 14 | 14.10 |

## 15.4 Using This Index in Practice

- **Educators** can assign inserts as standalone readings or thematic modules.
- **Consultants** can reference inserts in reports, proposals, or stakeholder workshops.
- **Publishers** can use the index for layout, citation, and digital tagging.
- **Students** can use the glossary and index to navigate the manuscript and build conceptual clarity.
- **Civic learners** can use glossary terms to understand planning decisions that affect their communities.

This chapter is not an appendix—it is a planning tool. It reflects your commitment to operational clarity, modular architecture, and ethical planning education.

Cities in Practice: First Edition 2025

www.ingramcontent.com/pod-product-compliance
Lightning Source LLC
LaVergne TN
LVHW090406210926
848480LV00002B/470